FOOD ALLERGIES: NEW RESEARCH

FOOD ALLERGIES: NEW RESEARCH

CARRIE M. CHESTERTON
EDITOR

Nova Science Publishers, Inc.
New York

For permission to use material from this book please contact us:
Telephone 631-231-7269; Fax 631-231-8175
Web Site: http://www.novapublishers.com

LIBRARY OF CONGRESS CATALOGING-IN-PUBLICATION DATA

Food allergies : new research / Carrie M. Chesterton, editor.
p. ; cm.
ISBN 978-1-60456-978-0 (hardcover)
1. Food allergy. I. Chesterton, Carrie M., 1959-
[DNLM: 1. Food Hypersensitivity. 2. Child. 3. Food--adverse effects. 4. Infant. WD 310 F6853 2008]
RC596.F645 2008
616.97'5--dc22 200803109

Published by Nova Science Publishers, Inc. ✦ New York

CONTENTS

PREFACE

A food allergy is an exaggerated immune response triggered by eggs, peanuts, milk, or some other specific food. Normally, your body's immune system defends against potentially harmful substances, such as bacteria, viruses, and toxins. In some people, an immune response is triggered by a substance that is generally harmless, such as a specific food. The cause of food allergies is not fully understood. A food allergy frequently starts in childhood, but it can begin at any age. Fortunately, many children will outgrow their allergy to milk, egg, wheat, and soy by the time they are 5 years old if they avoid the offending foods when they are young. Allergies to peanuts, tree nuts, and shellfish tend to be lifelong. This new book presents recent significant research in this field.

Short Communication - *Background and Objective*: Recently, the number of patients having oral allergy syndrome (OAS) associated with fruits has been increasing. We investigated the background, characteristics, and severity of the symptoms of such patients.

Subject and Methods: A questionnaire survey was conducted on the patients living in the Kanto region who visited the authors' hospital in the past 5 years and were suspected of having allergies to foods of plant origin.

Results: The subjects were 42 patients, 8 males and 34 females, whose mean age was 36 years. The complicating allergic diseases were allergic rhinitis in 35 patients (83%), asthma in 34 patients (81%), and atopic dermatitis in 14 patients (33%). The suspected causes of the OAS symptoms were rose family fruits in 31 patients, non-rose family fruits in 34 patients, vegetables in 14 patients, beans and nuts in 11 patients, and grains in 2 patients. As for the symptoms, oral and pharynx symptoms alone were observed in 12 patients, systemic symptoms were observed in 29 patients, and anaphylaxis was observed in 11 patients. Allergic rhinitis preceded OAS in 80% of the patients, which was a high incidence, but 20% of the patients did not experience this complication.

Conclusion: Rhinitic symptoms preceded OAS in many of the patients having oral allergies to foods of plant origin in the Kanto region. Moreover, the findings that there were patients who did not experience black alder pollinosis and that the symptoms were also caused at a high rate by non-rose family fruits suggests the presence of broad cross reactivity among pollens other those that of black alder and food of plant origin.

Chapter 1 - Food allergy (FA) represents one of the most important problems of pediatric allergology and immunology. FA diagnosing is very challenging and not as easy as it seems at first sight. It requires close cooperation among immunoallergologist, gastroenterologist, dermatologist and pediatrician. Diagnostic algorithm consists of detailed personal and family history, physical examination, tests for IgE mediated reactions (quantification of food-specific IgE and skin prick test), elimination diet and the "gold standard" – double-blind, placebo-controlled oral food challenge. However, all these methods and tests may be sometimes misleading. Additional test in the diagnosis of FA is atopy patch test (APT), which is aimed at the diagnosis of late clinical symptoms induced by special foods and can significantly contribute to the final outcome. The standardized method for APT testing has been published recently. APT has left experimental grounds and is increasingly used as a standard diagnostic procedure for characterizing patients with aeroallergen- and food-triggered disorders. Although APT seems a valuable additional tool in the diagnostic work-up of food allergy, especially in children with atopic eczema, the immunopathology and some technical aspects of testing remain controversial. There are still some points waiting to be answered, e.g., epidemiology of this tests in an unselected population, prevalence of side effects and safety of this test, possibility of sensitization through skin during testing, reproducibility in the same time and over-time, suitability of some new APT testing sets available on the market (e.g. plastic cups on tape), and especially optimal and good available testing substances.

For better understanding these unsolved aspects of this new diagnostic method, the authors performed the study in an unselected population of schoolchildren of two different nations (all together 900 children), where they examined the prevalence of positive APT with food and inhalant allergens. The authors evaluated the link between positive APT reactions and skin prick tests, circulating eosinophils, histamine skin reactivity, and questionnaire-derived frequencies of various atopic and non-atopic symptoms and diseases. The authors also investigated the right versus left and over-time reproducibility of duplicate APT with native and commercially available food and with inhalant allergens.

In this chapter, the authors review current knowledge about atopy patch test and to compare their results with already published studies. Despite some unresolved questions about APT, this test seems to be useful as an additional method in management of food-induced symptoms and disorders, but its results should be evaluated in the context with other methods and clinical status. Further studies are necessary for better understanding all the clinical characteristics and applications of atopy patch test.

Chapter 2 - Food allergy is defined as an adverse reaction because of an abnormal immunological response to food protein. The immune pathogenesis is, in the majority of cases, IgE-mediated although it may also be cell-mediated (non-IgE) or mixed IgE/cell-mediated. Food allergy affects as many as 2 to 8% of young children and the presentation can be highly variable. There is usually a clear temporal relationship between food exposure and the development of allergic symptoms. At times, symptoms may develop hours or days after food exposure making the diagnosis difficult.

Food allergy usually presents as multi-system involvement, most commonly gastrointestinal symptoms which occur with a frequency of 50 to 80% of cases. These are followed by cutaneous symptoms and respiratory symptoms, occurring in 20 to 40%, and 4 to 25% of cases, respectively. Gastrointestinal manifestations include oral allergy syndrome, gastrointestinal anaphylaxis, allergic eosinophilic esophagitis, allergic eosinophilic gastroenteropathy, food protein-induced enteropathy, food protein-induced enterocolitis syndrome, food protein-induced proctocolitis, gluten-sensitive enteropathy, infantile colic, irritable bowel syndrome, and constipation. Cutaneous manifestations are urticaria/angioedema, atopic dermatitis, contact dermatitis, and dermatitis herpetiformis. Rhinitis/rhinoconjunctivitis, asthma, Heiner syndrome, and serous otitis media are the respiratory manifestations of food allergy. Other manifestations include systemic anaphylaxis, food-dependent exercise-induced anaphylaxis, migraine, epilepsy, diabetes mellitus, nephrotic syndrome, nocturnal enuresis, anemia, thrombocytopenia, vasculitis, and arthropathy/arthritis. This chapter discusses the various clinical manifestations of food allergy.

Chapter 3 - Genuine food allergy affects approximately 5% of children and less than 1% of adults. The underlying mechanism is complex and involves immediate (type I) or delayed (type IV) sensitization to food proteins. The gold standard for the diagnosis of genuine food allergies is by double-blind placebo-controlled food challenge test. The literature gives ambiguous data on the association between food allergies and atopic diseases. In asthma, aeroallergens such as house dust, mites, and pollens are well known allergens. Apart from type I hypersensitivity reaction precipitating acute anaphylactic and asthmatic attacks by peanuts, egg or crustacean seafood, the association with food allergens is probably less prevalent. Allergic rhinitis/allergic rhinoconjunctivitis as the sole manifestation of food allergy is quite uncommon. Food allergy plays an immunopathogenic role in 30 to 50% of children with moderate to severe atopic dermatitis.

Chapter 4 - The definitive treatment of food allergy is strict elimination of the offending food from the diet. Symptomatic reactivity to food allergens is generally very specific, and patients rarely react to more than one food in a botanical or animal species. If elimination diets are prescribed, care must be taken to ensure that they are palatable and nutritionally adequate. Patients must have a good knowledge of food containing the allergen and must be taught to scrutinize the labels of all packaged food carefully.

Formula-fed infants with cow's milk allergy should be fed an elemental or extensively hydrolysed hypoallergenic formula. Soy formulas are inappropriate alternatives as a significant number of infants who are allergic to cow's milk are also allergic to soy. Most children outgrow their food hypersensitivity. As such, rechallenge testing for food allergy should be performed; the interval between rechallenges should be dictated by the specific food allergen in question, the age of the child, and the degree of difficulty in avoiding the food in question

Emergency treatment of food-induced anaphylaxis should follow the basic life support ABC principles, with the simultaneous intramuscular injection of adrenaline. A fast-acting H_1 antihistamine should be considered for the child with progressive or generalized urticaria or disturbing pruritus. Pharmacological therapies such as mast cell stabilizers have very little role to play in the treatment of gastrointestinal manifestations of food allergy.

In high-risk infants, exclusive breastfeeding with introduction of solid foods not earlier than 6 months of age may delay or possibly prevent the onset of food allergy in some children. Avoidance of allergenic foods by lactating mothers is often recommended. When breastfeeding is not possible, the use of a partially or extensively hydrolysed hypoallergenic formula is desirable. Prophylactic medications have not been shown to be consistently effective in the prevention of life-threatening reactions to food. Their use may mask a less severe reaction to a culprit food, knowledge of which might prevent a more severe reaction to that food in the future.

Chapter 5 - Infantile colic is characterized by paroxysms of uncontrollable crying or fussing in an otherwise healthy and well-fed infant less than 3 months of age. The duration of crying is more than 3 hours per day and more than 3 days per week for at least 3 weeks.

The condition can be very stressful to the family. The etiology is multifactorial. There is increasing evidence that cow's milk proteins may play an important role in the pathogenesis of infantile colic in a significant number of cases. Also, maternal ingestion of eggs, chocolate, citrus fruits, nuts, as well as certain seafood whilst breastfeeding may result in infantile colic. Intestinal permeability to macromolecules, a mechanism of acquired food allergy appears to be increased in some infants with colic. Supportive counseling, reassurance, and dietary modifications (if indicated) are the core measures used for the management of this condition. Use of hypoallergenic diets by breasting mothers should be considered at least for those infants with severe colic or with atopic features such as atopic dermatitis, asthma, and allergic rhinitis. For formula-fed infants with mild to moderate colic, the present consensus is that changing to another formula is usually not necessary. Formula-fed infants with severe colic, especially those with atopic features or a strong family history of atopy, may have a beneficial effect from hypoallergenic formulas such as whey hydrolysates or casein hydrolysates. Periodic challenges at monthly intervals are necessary to ensure that the improvement is related to dietary modification and not a result of natural resolution. In most infants, infantile colic resolves by 3 to 4 months of age.

Chapter 6 - Eosinophilic gastrointestinal disorders consist of diseases involving eosinophilic infiltration of the gastrointestinal tract. These include eosinophilic esophagitis (EE), eosinophilic gastroenteritis, and eosinophilic colitis. Much of the recent literature has described eosinophilic esophagitis with a relative lack of information regarding eosinophilic gastroenteritis and eosinophilic colitis specifically. Many studies of EE reviewed in this chapter included patients with extra-esophageal eosinophilic involvement. Much of what is known regarding EE is applicable to eosinophilic gastroenteritis, and perhaps, to a lesser extent, eosinophilic colitis. This chapter focuses mostly on eosinophilic esophagitis; however, the concepts are applicable to all the eosinophilic gastrointestinal disorders. Though first described decades ago, these entities are still, to some degree, poorly defined. As such, the diagnosis is described as clinicopathologic in that it relies on clinical signs and symptoms with supporting laboratory and histologic findings. As background, this chapter includes a brief description of the gastrointestinal tract barrier and oral tolerance. The gastrointestinal tract serves as a physical and immunological barrier. The normal response of oral tolerance reflects a lack of immunologic responsiveness as a result of prior exposure. Food hypersensitivity responses occur after a failure of oral tolerance development. This is important for all types of food hypersensitivity responses including the eosinophilic gastrointestinal disorders. Unlike immediate hypersensitivity (type I IgE-mediated) food allergic responses, eosinophilic esophagitis, eosinophilic gastroenteritis, and eosinophilic colitis may involve both IgE- and non-IgE-mediated responses. This is also in contrast to food protein-induced enterocolitis/colitis or celiac disease in which a cell-mediated mechanism is likely responsible without evidence of IgE involvement. In addition to the background information, this chapter describes the pathogenesis, clinical manifestations, as well as therapeutic approach to the eosinophilic gastrointestinal disorders.

Chapter 7 - The chapter reviews current research regarding the prevalence of severe food allergies in school age children and the most effective preventative and treatment practices to be implemented in schools. Common concerns revealed in the literature center around the risks involved when school personnel lack knowledge and awareness of anaphylaxis and the necessary emergency protocols to put in place in the event of a reaction. The authors examine the critical role of epinephrine in the early treatment of anaphylaxis and the urgent need for school nurses, physicians, parents, and educators to put effective protocols in place. To better ensure that epinephrine be administered promptly and that care is carefully coordinated with emergency personnel, school nurses need to be the coordinator of care. Analysis of recent studies regarding the school nurse's role in the development of emergency medical plans for children with special health care needs (CSHCN) reveals the need to have one full-time school nurse per school to better assure access to prompt, quality care. It is vital for school nurses to actively participate in the development of Individualized Education Plans (IEPs) or Individual Health Plans (IHPs) for students identified to have specific health related disabilities and be designated as the primary individual in charge of implementation of these plans. Across the United States there are dramatic differences in state and local policies impacting school nurse and student ratios which, in turn, can have a significant impact upon management and care provided to students with special medical needs. Recommendations for standards of care for students with severe food allergies will be discussed.

In: Food Allergies: New Research
Editor: Carrie M. Chesterton, pp. 1-7

ISBN 978-1-60456-978-0

Short Communication

CLINICAL FEATURES OF PATIENTS HAVING ORAL ALLERGY SYNDROME ASSOCIATED WITH PLANT FOOD ALLERGENS IN THE KANTO REGION

Emiko Ono*, Yuji Maeda, Hidenori Tanimoto, Yuma Fukutomi, Chiyako Oshikata, Kiyoshi Sekiya, Takahiro Tsuburai, Naomi Tsurikisawa, Mamoru Otomo, Masami Taniguchi, Toyota Ishii, Akihiko Asahina and Kazuo Akiyama

Clinical Research Center for Allergy and Rheumatology, Sagamihara National Hospital

ABSTRACT

Background and Objective: Recently, the number of patients having oral allergy syndrome (OAS) associated with fruits has been increasing. We investigated the background, characteristics, and severity of the symptoms of such patients.

Subject and Methods: A questionnaire survey was conducted on the patients living in the Kanto region who visited our hospital in the past 5 years and were suspected of having allergies to foods of plant origin.

Results: The subjects were 42 patients, 8 males and 34 females, whose mean age was 36 years. The complicating allergic diseases were allergic rhinitis in 35 patients (83%), asthma in 34 patients (81%), and atopic dermatitis in 14 patients (33%). The suspected causes of the OAS symptoms were rose family fruits in 31 patients, non-rose family fruits in 34 patients, vegetables in 14 patients, beans and nuts in 11 patients, and grains in 2 patients. As for the symptoms, oral and pharynx symptoms alone were observed in 12 patients, systemic symptoms were observed in 29 patients, and anaphylaxis was observed in 11 patients. Allergic rhinitis preceded OAS in 80% of the patients, which was a high incidence, but 20% of the patients did not experience this complication.

* All correspondence and reprint requests should be addressed to: Emiko Ono; Clinical Research Center for Allergy and Rheumatology, Sagamihara National Hospital, Sakuradai 18-1, Sagamihara, Kanagawa 228-8522, Japan Tel: +81-42-742-8311, Fax: +81-42-742-5314, E-mail: e-ono@sagamihara-hosp.gr.jp

Conclusion: Rhinitic symptoms preceded OAS in many of the patients having oral allergies to foods of plant origin in the Kanto region. Moreover, the findings that there were patients who did not experience black alder pollinosis and that the symptoms were also caused at a high rate by non-rose family fruits suggests the presence of broad cross reactivity among pollens other those that of black alder and food of plant origin.

INTRODUCTION

Recently, an increase in the number of cases of allergy to fruits and vegetables in adults has been reported[1]. Because some types of fruit and vegetable allergies develop in association with pollinosis, these allergies are called pollen-food allergy syndrome (PFAS) or oral allergy syndrome (OAS). OAS caused by rose family fruits in association with birch pollinosis is well known in Europe, and the OAS caused by fruits has also been reported, mainly in Hokkaido in Japan where birch grows naturally. However, there have been only a few reports on the actual condition of OAS patients in the Kanto region where, different from Hokkaido, birch does not grow naturally. The purpose of this study was to investigate the frequencies of OAS in the Kanto region and to compare them with previous reports.

SUBJECTS AND METHOD

Subjects

The subjects were 42 patients who visited our hospital (Department of Allergy, Otolaryngology and Dermatology) from 2000 to 2005 and were considered to show allergy syndromes caused by food of plant origin on the basis of their history of the disease. The designation OAS was used regardless of the severity of the disease in this study.

Methods

A survey was conducted on the patients who mentioned during an interview on their first visit that they experienced allergy symptoms because of fruits. On the basis of that survey, their allergy symptoms and possible causative foods were evaluated. A prick test was performed in some cases. The causative food was surmised on the basis of their history of the disease regardless of the results of the radioallergosorbent test (RAST) for specific IgE antibodies and the prick test. The main questions were as follows:

1. Type of causative food (rose family fruits, non-rose family fruits, vegetables, grains, and nuts)
2. Expression frequency of each symptom
3. Age at onset
4. Amount of intake and time to appearance of symptoms
5. Presence of symptoms caused by processed food and heated food
6. Related factors (including physical condition, drugs being taken, and exercise)

7. Presence of complicating allergic disease
8. Relationship between black alder pollinosis and rose family fruits
9. Temporal relationship between onset of rhinitis and that of OAS

RESULTS

Background of Patients (Table 1)

Table 1 shows the background of the patients. The mean age of the 42 patients was 36 years. Sixty percent or more of the patients were female, showing the tendency that young women are more prone to the disease. As for complications of other allergic diseases, 34 patients (81%) among the 42 patients had bronchial asthma, 35 (83%) had allergic rhinitis, and 14 (33%) had atopic dermatitis. While the main subjects in the conventional OAS reports were pollinosis patients, the subjects in this study were OAS patients. Therefore, the rate of complication of allergic diseases in this study also differed from that in the conventional OAS reports.

1) Type of Causative Food (Figure 1 and Figure 2)

The most common causative food was fruits, then vegetables, beans, and grains, in that order. The number of patients who had OAS caused by rose family and non-rose family fruits was almost identical. Peach, apple, pear, and strawberry, in that order, were common causative fruits of the rose family. Melon, kiwi, orange, and pineapple, in that order, were common causative fruits of the non-rose family (Figs. 1 and 2).

2) Expression Frequency of Each Symptom (Figure 3)

Oral symptoms, which seem to be caused by direct contact with the allergen, were the most common symptoms observed in 92.9% of the patients. Skin symptoms were observed in 57.1% of the patients, gastrointestinal symptoms in 45.2%, conjunctivitic and rhinitic symptoms due to pollinosis in approximately 25%, and respiratory symptoms in 14%. Approximately 30%, a high rate, of the patients experienced anaphylaxis. As for the details of the oral symptoms, itching, swelling, and discomfort of the pharynx were observed in almost all the patients, and hoarse voice suggesting edema of the pharynx or vocal cord was observed in approximately 10% of the patients. Hives, angioedema, and cutaneous pruritus were common skin symptoms, and abdominal pain and nausea were common gastrointestinal symptoms.

3) Amount of Intake and Time to Appearance of Symptoms

The amount of intake was one bite or less in approximately 70% of the patients. Symptoms appeared in the mouth, lips, and pharyngeal mucosa, which had direct contact with the causative food, immediately after the intake in most of the patients. As for the time to the appearance of symptoms, the most common answer was "immediately after intake," and the symptoms appeared within 10 minutes or less in 80% of the patients.

4) Presence of Symptoms Caused by Processed Food and Heated Food

Allergic symptoms were caused by processed food including heated food in approximately 30% of the patients. The symptoms were caused by heated food in 7 patients (17%).

5) Related Factors (Including Physical Condition, Drugs Being Taken, and Exercise)

The symptoms appeared in approximately 70% of the patients in normal physical conditions. The symptoms seemed to be associated with poor physical condition, such as fatigue or having the common cold, in approximately 20% of the patients.

6) Presence of Complicating Allergic Disease

The rate of complications of asthma and allergic rhinitis was high (Table 1).

7) Relationship between Black Alder Pollinosis and Rose Family Fruits

Nineteen patients were presumed to have black alder pollinosis, and 12 patients had pollinosis other than that of black alder. Black alder pollinosis was diagnosed on the basis of the time of appearance of symptoms and the results of the intradermal test or RAST. Eighty-four percent of the black alder pollinosis patients had allergies to rose family fruits, which was high. However, 58% of the patients having pollinosis other than that of black alder had allergies to rose family fruits, which was also high. There was no significant difference between the two groups of patients. This issue requires further study on a larger number of subjects.

8) Temporal Relationship between Onset of Rhinitis and of OAS

As for the temporal relationship between the onset of rhinitis and that of OAS, rhinitic symptoms preceded the onset of OAS at a very high rate. The number of cases in which OAS preceded rhinitic symptoms was only 3 out of 42. Rhinitic symptoms occurred within 3 years after the onset of OAS in these 3 patients; the onset of OAS and that of rhinitic symptoms were close. This result indicates a close relationship between the onset of OAS and that of rhinitis.

Discussion

The disease name, oral allergy syndrome (OAS), was first proposed by Amlot et al. [2] in the UK in 1987. Ortolani et al. [3] demonstrated the relationship between OAS and pollinosis, particularly birch pollinosis, the following year. Since then, OAS has been considered as an immediate-type allergic reaction caused by antigenic molecules that are common to pollen allergens and plant food allergens. Type II food allergy, to which OAS belongs, is characterized by differences between the route of allergenic sensitization and occurrence, and between the sensitizing antigen and the eliciting antigen [4]. It has been reported that many fruits and vegetables that cause OAS show cross reactivity with pollen. The pollen of the birch family has been known to have cross reactivity with black alder pollen [5]. Birch grows in Hokkaido and in the highlands of Honshu in Japan. On Honshu, birch grows only in a particular area such as Nagano. Therefore, some kind of pollen other than that of birch is considered to be the causative pollen on Honshu. It has been reported that the black alder pollen in the Kanto regions and the pollen of oba-yashabushi (*Alnus sieboldiana*), in addition to the black alder pollen, in part of the Kansai region are the major pollens associated with OAS [6,7]. The main allergen of the pollen of oba-yashabushi and that of the black alder have been reported to have a highly common antigenicity with Bet v 1 of birch pollen [6].

Eighty percent of the OAS patients investigated in this study showed rhinitis complications, and the onset of allergic rhinitis preceded that of OAS in 90% of those patients. Because black alder pollen is dispersed in May and June when the cedar pollen season ends in the Kanto regions, we cannot deny the possibility that some OAS patients investigated in this study were suffered from pollinosis caused by pollen other than that of black alder, such as the pollen of orchard grass. On the basis of the positive reaction in the prick test or RAST and rhinitis symptoms observed in May and June, 19 patients among the subjects in this study were diagnosed as having black alder pollinosis. These patients had a high rate of allergies to rose family fruits (16 out of 19 patients (84%)), and patients who did not have black alder pollinosis also had allergies to rose family fruits at a high rate (8 out of 12 patients (67%)). It was presumed from these results that OAS caused by rose family fruits is also associated with pollen other than that of black alder, namely, pollen other than that of the birch tree family. It has been reported that the OAS caused by melon, watermelon, and peach is often observed in patients with double sensitization to birch pollen and gramineous plants, and that the OAS caused by melon is often observed in patients with double sensitization to birch pollen and mugwort [8,9]. The relationship between OAS and the causative antigen and the positive rate of specific IgE to the inhaled antigen in the pollinosis patients, and pollinosis other than birch pollinosis, should be further examined and compared with the results of this study.

OAS is well known, and there have been reports of it from Hokkaido [8-12], Yamanashi [13,14], and the Kansai regions [6,7]. However, the details of the clinical features of OAS patients in Japan are unclear. We compared the results of our study on the clinical features of OAS patients in the Kanto region with those previously reported. It was reported that rose family fruits, particularly apple, peach, and cherry, are the common fruits causing OAS symptoms in birch pollinosis patients [8]. Matsuzaki et al. reported that rose family fruits are also the common causative fruits in Yamanashi [14]. Because the subjects of that report were patients having cedar and cypress pollinoses, it was unclear as to what extent the black alder pollinosis patients had OAS complications. In terms of plant systematics, the pollens of cedar and cypress do not show common antigenicity with that of birch family trees according to previous reports. The results that peach, of the rose family, and melon, kiwi, and pineapple, of the non-rose family, are the common causative fruits were similar to the results of our study. Naturally, the OAS frequency tended to be high if the causative fruits were consumed in large amounts in the study area. The time from intake to the appearance of symptoms was within 15 minutes in 85% of the patients, which was almost the same as that in our report [11,14]. As for the gender of the patients, although some researchers have reported that there is no gender difference in the frequency of OAS, recently, there have been many reports indicating that the frequency of OAS is high in women [8,12,14,15]. The frequency of OAS was markedly high in women in this report. Considering that OAS caused by fruits occurs together with pollinosis and that pollinosis is more common in women, the result indicating the high frequency of OAS in women seems to be reasonable. As for the symptoms, the number of mild cases with oral symptoms alone in our investigation was smaller than that in previous reports. In most of the previous reports on OAS, the subjects were pollinosis patients, and the frequencies of complication of pollinosis with OAS were investigated. The percentage of severe OAS patients in our study and that in the previous reports seemed to differ because of the difference in the type of subjects. In this study, we did not limit the subjects to pollinosis patients but also investigated patients who showed some allergic symptoms to food of plant origin. Therefore, the results obtained in this study do not show the clinical features of OAS patients having a certain type of pollinosis. Kato investigated the rate of complication of OAS in each underlying disease and reported that the rate of the complication of OAS was approximately 30% in the patients who had both atopic dermatitis and asthma [15]. We should seriously consider the possibility that the difference in the subjects in each study affected some of the results. The difference in the subjects affects the complication rate of allergic diseases other than pollinosis as well as the frequency and severity of OAS. The reason for the high rate of severe cases in the subjects of this study is unclear. We will classify the subjects according to the severity of their symptoms and reexamine the clinical background and characteristics of the patients with severe symptoms in our second report.

Conclusion

In the Kanto region, 80% of the OAS patients showed rhinitis complication, and the rhinitis symptoms preceded the onset of OAS in 90% of such patients. The result indicates that the OAS of the patients in the Kanto region is strongly related to pollen sensitization and thus belongs to Type II food allergy similar to the OAS of the patients with birch pollinosis in Hokkaido. The results that there were OAS patients without the black alder pollinosis complication in the Kanto region, that the patients showed allergic reaction to the non-rose family fruits with almost the same frequency as to the rose family fruits, and that the rhinitis patients other than the black alder pollinosis patients also had OAS symptoms attributable to rose family fruits with a high rate suggested a broad cross reactivity with pollens of trees and weeds. The issue of the cross reactivity, including the relationship of OAS with pollinosis other than that from the birch family of trees, should be further examined.

References

[1] Zenro Ikezawa. Oral allergy syndrome (OAS) and allergies to fruits and vegetables. *Allergology* 2004; 18: 537-545.

[2] Amlot PL, Kemeny DM, Zachary C et al. Oral allergy syndrome: symptoms of IgE mediated hypersensitivity to foods. *Clin Allergy* 1987; 17: 33.

[3] Ortolani C, Ispano M, Pastorella EA, et al. Comparison of results of skin prick test (with fresh foods and commercial food extracts) and RAST in 100 patients with oral allergy syndrome. *J Allergy Clin Immunol* 1989; 83: 683.

[4] Takeshi Yagami. Similarities between latex-fruits syndrome and oral allergy syndrome associated with pollinosis. *The Allergy in Practice* 2000; 20: 854.

[5] Hoffmann SK, Susani M, Ferreira F, et al. High level expression and purification of the major birch pollen allergen *Bet V1.Protein Expression Purif* 1997; 9: 33.
[6] Shiro Yoshimura. Oba-yashabushi (*Alnus sieboldiana*) pollinosis and OAS. *Igaku No Ayumi* (*Journal of Clinical and Experimental Medicine*) 2004; 209: 155-159.
[7] Atsuko Adachi and Tatsuya Horikawa. Regional difference in oral allergy syndrome associated with pollinosis. *Japanese Journal of Allergology* 2006; 55: 811-819.
[8] Tetsuo Yamamoto, Kohji Asakura, Hideaki Shirasaki, Tetsuo Himi, Hideki Ogasawara, Shin-ichiro Narita and Akikatsu Kataura. Sensitization to pollen and oral allergy syndrome. *Journal of Otolaryngology of Japan* 2005; 108: 971-979.
[9] Tetsuo Yamamoto, Kohji Asakura, Hideaki Shirasaki, Tetsuo Himi, Hideki Ogasawara, Shin-ichiro Narita and Akikatsu Kataura. Birch pollinosis in Sapporo and oral allergy syndrome. *Japanese Journal of Allergology* 2004; 53: 435-442.
[10] Kohji Asakura and Tetsuo Yamamoto. Pollinosis in Hokkaido region and oral allergy syndrome. *Allergology and Immunology* 2001; 8: 846-851.
[11] Megumi Kumai. Birch pollinosis and oral allergy syndrome in our hospital. *Stomato-pharyngology* 2001; 13: 179-188.
[12] Takuro Tosho and Shin-ichi Kawabori. Birch pollinosis and oral allergy syndrome. *Stomato-pharmagology* 2001; 13: 209-214.
[13] Zensei Matsuzaki and Yoshitaka Okamoto. Pollinosis in Kanto region and oral allergy syndrome. *Allergology and Immunology* 2001; 8: 852-857.
[14] Zensei Matsuzaki and Tomokazu Matsuoka. Oral allergy syndrome in questionnaire for patients with cedar and cypress pollinosis. *Journal of Japan Society of Immunology and Allergology in Otolaryngology* 2003; 21: 66-67.
[15] Yukihiko Kato. Clinical condition of skin disease, oral allergy syndrome (OAS) and pollen. *Japanese Journal of Clinical Dermatology* 2001; 55: 33-36.

In: Food Allergies: New Research
Editor: Carrie M. Chesterton, pp. 9-90

ISBN 978-1-60456-978-0

Chapter 1

CURRENT POSITION OF ATOPY PATCH TEST IN THE DIAGNOSIS OF FOOD ALLERGY IN CHILDREN

Milos Jesenak[1,2*], Zuzana Rennerova[1,3], Eva Babusikova[4], Peter Banovcin[2], Lubica Jakusova[2], Zuzana Havlicekova[2], Mario Barreto[1], Maria Pia Villa[1] and Roberto Ronchetti[1]

[1]Department of Pediatrics, 2nd School of Medicine,
University "La Sapienza", Rome, Italy
[2]Department of Pediatrics, Jessenius School of Medicine,
Comenius University in Bratislava, Martin, Slovakia
[3]Pneumo-Alergo Centrum, Bratislava, Slovakia
[4]Institute of Medical Biochemistry, Jessenius School of Medicine,
Comenius University in Bratislava, Martin, Slovakia

"Anyone can do a patch test.
Few do it well.
Fewer can properly interpret patch tests."

–Albert Kligman

ABSTRACT

Food allergy (FA) represents one of the most important problems of pediatric allergology and immunology. FA diagnosing is very challenging and not as easy as it seems at first sight. It requires close cooperation among immunoallergologist, gastroenterologist, dermatologist and pediatrician. Diagnostic algorithm consists of detailed personal and family history, physical examination, tests for IgE mediated

* Milos Jesenak, M.D., Ph.D.; Clinica Pediatrica, Ospedale Sant'Andrea, Via Grottarossa 1035/1039, 00189 Rome (RM), Italy
Department of Pediatrics, Comenius University in Bratislava, Jessenius School of Medicine, Kollarova 2, 036 59 Martin, Slovak Republic
Tel.: 00421-903-273-531, Fax: 00421-043-422-2678; Mail: jesenak@gmail.com

reactions (quantification of food-specific IgE and skin prick test), elimination diet and the "gold standard" – double-blind, placebo-controlled oral food challenge. However, all these methods and tests may be sometimes misleading. Additional test in the diagnosis of FA is atopy patch test (APT), which is aimed at the diagnosis of late clinical symptoms induced by special foods and can significantly contribute to the final outcome. The standardized method for APT testing has been published recently. APT has left experimental grounds and is increasingly used as a standard diagnostic procedure for characterizing patients with aeroallergen- and food-triggered disorders. Although APT seems a valuable additional tool in the diagnostic work-up of food allergy, especially in children with atopic eczema, the immunopathology and some technical aspects of testing remain controversial. There are still some points waiting to be answered, e.g., epidemiology of this tests in an unselected population, prevalence of side effects and safety of this test, possibility of sensitization through skin during testing, reproducibility in the same time and over-time, suitability of some new APT testing sets available on the market (e.g., plastic cups on tape), and especially optimal and good available testing substances.

For better understanding these unsolved aspects of this new diagnostic method, we performed the study in an unselected population of schoolchildren of two different nations (all together 900 children), where we examined the prevalence of positive APT with food and inhalant allergens. We evaluated the link between positive APT reactions and skin prick tests, circulating eosinophils, histamine skin reactivity, and questionnaire-derived frequencies of various atopic and non-atopic symptoms and diseases. We also investigated the right versus left and over-time reproducibility of duplicate APT with native and commercially available food and with inhalant allergens.

In this chapter, we would like to review current knowledge about atopy patch test and to compare our results with already published studies. Despite some unresolved questions about APT, this test seems to be useful as an additional method in management of food-induced symptoms and disorders, but its results should be evaluated in the context with other methods and clinical status. Further studies are necessary for better understanding all the clinical characteristics and applications of atopy patch test.

1. Introduction

Food hypersensitivity (FH) affects nearly everyone at same point, either as an unpleasant reaction to something eaten or as a concern for a family member suspected of having food allergy (FA). Increasing public and medical interest have also popularized claims that a variety of physical and psychological symptoms are the result of food hypersensitivity. The ratio of perceived FH in the general population is about 25%. However, the prevalence of true food allergy is approximately 8% in children under 1 year, 2-3% in children between 1-3 years and about 1-2% in adults [3, 188]. Therefore, it is very important to determine if the food really causes these symptoms or if there are other underlying factors. According to the revised classification of allergic diseases, food hypersensitivity can be divided into two groups: immunologically-mediated reactions (**food allergy**) and non-immunologically-mediated reaction (**food intolerance**). According to the involved immune mechanism, food allergy can be divided into **IgE-mediated** (reaction type-I), **non-IgE-mediated** (cellular FA, reaction type III or IV) and **combined** (mixed IgE- and non-IgE-mediated) reactions [96]. Non-IgE mediated food allergy is supposed to be a cell-mediated immunologic reaction, which involves immune complex formation and complement deposition [198]. Non-IgE-mediated food sensitivities are becoming increasingly recognized. This group is represented by a spectrum of clinical diseases attributed to adverse immune responses to food for which IgE antibodies to the causal food can not be demonstrated, at least not by routine tests. The onset of these reactions is slower than immediate IgE-mediated reactions, ranging from a few hours to more than a week after the ingestion of the causative agent [152]. In some cases, even more prolonged and repeated exposure is required for the development of clinically apparent abnormalities. Most of the diagnosis is made on the basis of clinical presentation and response to dietary exclusion (dechallenge) [3]. IgE-mediated sensitization can be shown in the large majority of cases with proven FA. However, in 10% of children with atopic dermatitis (AD) and positive oral food challenge (OFC), both the skin prick test (SPT) and food-specific IgE (sIgE) measurement are negative [146]. Identification of the causal foods for non-IgE-mediated disorders is complicated. The symptoms are often subacute or chronic in nature and no simple tests are available to secure the diagnosis.

Early identification of children who would profit from strict avoidance of specific dietary allergens is important especially for:

- to avoid unnecessary elimination diets involving risk of growth retardation in early periods of life,
- to ameliorate the clinical course of AD,
- as secondary prevention of the development of multiple food allergies or respiratory allergic diseases [92].

Food allergy represents one of the most important problems of pediatric allergology and immunology. **Diagnosing of FA** is very challenging and not as easy as it seems at first sight. In the correct diagnosis of FA close cooperation among immunologists, allergologist, gastroenterologist, dermatologist and pediatrician is necessary. As in other diseases, the allergy diagnosis is established in several steps, with detailed **analysis of personal and family history** and **careful physical examination** at the beginning. **Diet diaries** are used as adjunct to history over a specified time period and may help to reveal unknown sources of food allergens. **Skin prick test** with native foods (so-called prick-by-prick technique) may be more reliable than SPT with commercially extracts in screening patients with suspected IgE-mediated FA. While negative SPT, according to some authors, nearly exclude IgE-mediated allergy, positive tests do not prove relevant allergy. **Intradermal skin tests** have no higher predictive value, but bear a higher risk of systemic reactions. In few studies, **food-specific IgE** to food allergens showed good correlation with provocation test results [238]. However, it is the only method in patients with dermographism, severe eczema or antihistamine medication and has its role in the general concept of allergy diagnosis [186]. **Specific elimination diets** should be initiated before **oral exposition test**, which remains "the gold standard" in the diagnosis of FA. Preferably, the provocation is performed as **double-blind placebo controlled food challenge** (DBPCFC), e.g. with masked (lyophilized) foods in colored and flavored neutral formulas after at least 2 weeks of corresponding elimination diet [40]. Nevertheless, this method is time consuming, costly, brings the risk of severe systemic anaphylactic reactions and must be performed over several days under control of physicians [158]. The cornerstone of FA diagnosis is relief of symptoms on elimination of the suspected food stuff and the return of symptoms on reintroduction [184]. A lot of studies tried to develop some alternative test to make DBPCFC, at least in some cases, superfluous.

Many laboratory tests have been introduced, but their diagnostic value is limited by several weaknesses; no single test is able to identify all the patients with FA. This is probably because several immunological mechanisms are involved in FA. The age of the tested subject may influence the test results, and allergens triggering the symptoms may vary between individuals [180]. Only the IgE-mediated reaction, usually associated with an immediate-type reaction, is well characterized, the immunological mechanisms associated with a delayed-type reaction are still poorly understood. Though many children with atopic dermatitis are positive with food allergens in the skin prick testing, an index of immediate-type reaction, the delayed-type reaction tests, effective for determining whether food allergens could by involved in the atopic dermatitis or not, are not routinely performed [135].

Various kinds of skin testing are a common diagnostic procedure in a management of food allergy, but the final diagnosis of FA is based on the clinical response to the food exposition in food challenge [54]. Among these tests, the most frequently used skin tests are:

- **Skin prick test**: this test investigates type I sensitization, the IgE-mediated reactions,
- **Skin scratch test**: the clinical use of this test was abandoned because of poor reproducibility,
- **Intradermal test**: this test has not been proven as reliable and brings high risk of systemic reactions,
- **Skin application food test**: this test investigates IgE-mediated acute contact urticaria released by an epidermal application of allergens, but seems to be not reproducible enough, and finally,
- **Atopy patch test**: newly introduced test aimed to detect type IV reaction (T-cells-mediated reactions).

The diagnostic reliability of skin tests for FA depends on several factors including patient selection, symptoms considered, standardization of the allergen extract, and kind of food involved [160, 185]. Skin testing is a common diagnostic procedure in FA, but the final diagnosis is based on the clinical response to food. The only accepted methods for verifying the diagnosis of FA are still the double-blind, placebo controlled, food challenge and open controlled food challenges in young children [21]. Today, different diagnostic *in vivo* and *in vitro* tests are available, but their diagnostic accuracy are not sufficient making them diagnostic tools, usually because of poor quality of the extracts used or depending on the technical aspects of the test. The use of fresh foods for testing has been shown to improve sensitivity and specificity of studied tests compared with the results of DBPCFC [148]. A positive outcome of the tests (after exclusion of false positive results) indicates sensitization, but does not mean that exposition to these allergens can cause clinical allergic symptoms. The final diagnosis should be definitely established according to the results of controlled elimination diet and oral food challenges [21]. Early recognition of dietary allergies in infants is essential for avoidance of unnecessary diets, amelioration of skin disease, and secondary prevention of the development of multiple food allergies or respiratory allergic diseases [92].

2. Food Allergy and Atopic Dermatitis

Atopic dermatitis (AD) is a clinically defined, highly pruritic skin disease frequently associated with IgE responses against various allergens. AD is multifactorial, chronically relapsing inflammatory skin disorder which is characterized by erythematous and pruritic lesions, excoriations, papules and lichenification [78]. Aeroallergens and food allergens are the most important among the allergens found to be relevant in atopic eczema [58]. Food allergens play an important role in the exacerbation, occurrence and maintenance of skin symptoms in children with AD [164]. Besides the chronic skin manifestation of AD, which is usually due to food ingestion, various kinds of food are capable of causing one or more types of contact dermatitis. Contact allergies to foods, spices, and food additives can occur to individuals in the workplace or at home. Six different reactions have been described: irritant contact dermatitis, allergic contact dermatitis, contact urticaria, phototoxic contact dermatitis, photo-allergic contact dermatitis, and systemic contact dermatitis [13]. The skin exposition tests are also important in the diagnosis of these disorders (skin prick tests, patch tests, atopy patch tests and skin application food tests).

There is increasing consensus about the significance of food allergens in the pathogenesis and in the induction of atopic dermatitis in infancy and childhood, with cow's milk (CM) and hen's egg (HE) accounting for the most reactions. Hen's egg, cow's milk, soy and wheat account for about 90% of allergenic foods in children with AD [139]. Clinical relevant food hypersensitivity has been demonstrated in 30-70% patients with mild to severe AD according to the results of DBPCFC [28, 108, 187-188].

Basically, there three kinds of cutaneous reactions in AD patients which can be observed in connection with food [244]:

- Immediate-type reactions such as urticaria, angioedema and erythema occurring a few minutes after ingestion of food without an exacerbation of AD. Additionally to these reactions also gastrointestinal, respiratory or cardiovascular symptoms can be seen.
- Pruritus occurring soon after the ingestion of food with subsequent scratching leading to an exacerbation of AD.
- Exacerbations of AD occurring after 6-48 hours (delayed-type reactions). These reactions can occur also after an immediate-type response.

Identification of offending allergens, food or inhalant, is very important in the management of AD. Accurate and objective demonstration of a causal relationship between the dietary allergens and exacerbation of AD allows for the compliance of the family to the treatment, is the condition of growth in early life and is essential for avoidance of unnecessary elimination diets. Confirmation of the diagnosis and adequate management plays the role also in the secondary prevention of multiple food allergies development and bronchial asthma [32]. Children in whom AD does not improve despite routine treatment with emollients and topical corticosteroids or immunomodulators should be tested for allergy to foods, particular CM [92, 187].

The diagnosis of food allergy in AD children is especially clinical and relies on the disappearance of the symptoms during an elimination diet (dechallenge) and their subsequent reappearance after reintroduction of the food into the diet. IgE-mediated reactions are responsible for immediate reactions and cell-mediated reactions have been proposed to be responsible for delayed reactions. The mechanisms of delayed-type clinical reaction are not yet clearly understood and there is no suitable diagnostic tool for revealing the causal food allergen responsible for provoking of these reactions. While immediate-type clinical reactions during oral food challenge with food allergens can easily be identified, the evaluation of a possible FA in the absence of immediate clinical reactions still presents diagnostic difficulties [28, 108, 187-188]. Positive SPT and/or elevated food-specific IgE in serum traditionally demonstrate an IgE-mechanism. Most food allergies involve an IgE-mediated hypersensitivity reaction, and patients with AD very often show immediate reactions in skin tests with food allergens. Positive SPT with food allergens, however, do not necessarily imply a role for these allergens in the pathogenesis of AD, since oral food challenges demonstrated that in a subset of patients with AD, acute-onset clinical reactions can be observed, whereas the others had delayed-onset eczematous reactions [92]. No relationship has been shown between the reactivity in SPT and delayed-onset clinical reactions [246]. In the case of delayed hypersensitivity, the causal relationship between the ingestion of particular food and the appearance of symptoms is more difficult to detect, since the symptoms may appear from several hours to several days after the exposition to the causal antigen [93]. Atopy patch test is newly introduced possibility for detecting the delayed reactions characterized by T-cells activation responsible for those late-onset reactions in AD or other disorders connected with FA. Studies of the APT with particular allergens (CM, HE, Soy, wheat) in children have shown improved utility for determining responses to oral food challenges [92, 103, 123, 124, 140, 172, 180]. APT may become an important diagnostic tool, especially in patients with non–atopic AD, where the standard SPT and blood analyses do not identify those allergens which are clinically relevant for the course of this disease [74]. Atopy patch test is defined as an epicutaneous patch test with allergens known to elicit IgE-mediated sensitization and the evaluation of these eczematous skin reactions. IgE-associated activation of allergen specific T-cells by the food allergens, applied on the skin surface, probably via the activation of Langerhans cells, can lead to the elicitation of eczematous skin reaction in APT.

3. Atopy Patch Test – Introduction

A simple, unexpensive and reliable test for food allergy has been sought by food allergists for decades. Recently, the atopy patch test has been introduced into clinical use. APT (atopy patch test, atopic patch test, skin patch test, epicutaneous patch test, allergen patch test, allergic patch test, cutaneous contact test) has left experimental grounds and is increasingly used as a standard diagnostic procedure for characterizing patients with aeroallergen- and food-triggered disorders. Because the clinical morphology of AD corresponds to eczema and not to urticaria with wheal- and flare-type reaction, it was necessary to search a test procedure preferentially addressing the cellular, eczema-type component of the atopic skin reaction. The test procedure of APT is very similar to the classic patch test; it differs in the nature of allergens used. These are not, as in classic patch test, haptens, but intact protein allergens that are frequently used for SPT to demonstrate and IgE-mediated type I sensitization. Hence, the APT is not only a delayed-type hypersensitivity reaction as initiated in tuberculosis skin test [105]. In the APT we supposed the predominance of cellular-mediated reaction with evident participation of other immunological mechanisms (IgE, IgG, circulating immunocomplexes, complement, etc.). The APT has been used as a model for early AD lesions and is performed like a normal patch test with haptens, only with protein allergens. The atopy patch test could be defined as an epicutaneous patch test known to elicit IgE-mediated reactions, in which the test sites are evaluated for an eczematous reaction after 48 to 72 hours [170]. Primarily, the APTs were studied for aeroallergens, but several years ago, the APT with food allergens were introduced into clinical practice.

In the history of APT, there are several important milestones:

- **1937** – Rostenberg and Sulzberger [177] published first experimental study describing patch testing with aeroallergens in provocation of eczematous skin lesions. This was the earliest publication on patch testing in eczema.
- **1976** – A Japanese group [221] reported positive patch test to human dander.
- **1982** – Mitchell et al. [133] described eczematous reactions after patch testing with well-defined aeroallergens. They demonstrated that epicutaneous application of several allergens on the uninvolved, abraded skin of the patients with severe AD could induce eczematous lesions only in patients who also showed a positive immediate skin reaction to the same allergen. This was the earliest clinical controlled trial.
- **1989** – Ring et al. [170] named this epicutaneous patch test with allergens known to induce IgE-mediated sensitization as "atopy patch test"..
- **1989** – Breneman et al. [16] as first used food patch test in the group of 400 patients with diagnosed food allergy. They studied immunopathological changes in the skin after epicutaneous application of many food allergens.
- **1996** – Kekki et al. [103] and Isolauri et al. [92] showed that APT has a role in food allergy diagnostic work-up, especially in children suffering from AD.

Thereafter, many groups have studied APT with aeroallergens and food allergens. The outcomes of these studies show large variations, due to differences in patient selection, and more importantly, differences in methodology [49, 111]. More than 100 articles have been published on the APT technique and performance with many suggested improvements in vehicles and allergen concentrations [34, 105, 143, 145, 206] which led to the Position paper of European Academy of Allergy and Clinical Immunology in 2006 on Atopy patch testing [220]. The major disadvantage of patch testing is that it is less reproducible than other *in vivo* tests because of missing standardization of applied allergen and because of difficulties in differentiating irritative reactions from true allergic responses [79]. The variables in the outcomes of published studies are due to many important aspects and modifications in the APT technique (fig. 1) [2].

4. Skin Application Food Test

Skin application food test (SAFT) could be considered as a forerunner of today widely-used atopy patch tests with food allergens. Acute skin manifestation of IgE-mediated FA is called contact urticaria (contact urticaria syndrome) which can be combined with AD. For diagnosis of these reactions, a provocation test system, so-called skin application food test has been developed [147, 215]. In SAFT, food extracts were applied with Finn Chambers on volar surface of the forearm only for a few minutes and the results were read after 15-30 minutes. SAFT was based on the mechanism of contact urticaria being a patch test applied for only several minutes. It was supposed to be useful in detecting early phase type I immediate hypersensitivity reactions [157]. It was especially used in children younger than 3 years. The fresh food extract was applied on the volar surface of the forearm using 12 mm aluminium Finn Chambers (the same set as for atopy patch test) and the test was read after 15-30 min. The results were scored as follows: 1 = no reaction, 2 = erythema, 3 = erythema and oedema within the area of testing chamber, 4 = erythema and oedema overlying the borders of the chamber. A score more than 3 was regarded as positive [51].

Sources of Variability in Atopy Patch Testing

Materials	Methodology	Biological aspects
type of atopy patch test (delivery system)	criteria of patient´s selection	regional variation in skin responsiveness
different sources of APT allergens	control groups of healthy volunteers in the studies	regional variation in skin absorption
different vehicles of APT allergens	occlusion and reading time	weak and doubtful responses
different vehicles and concentrations of allergens	errors in the sequence of consecutive allergens	summation of individual responses
variation of allergen content in fresh food stuff	dissimilar pressure supported by the system	cyclic variation throughout the menstrual cycle
uneven distribution of allergen content in vehicle	partial or complete detachment of the patches	systemic and topical Medications withdrawal
deposition of allergen in the APT device	sweating	skin hyporactivity and „silent back syndrome"
imperfect occlusion of the chambers to the skin	interpretation of the results (scoring scales)	skin hyperractivity and „excited skin syndrome"

Ale IS et al. Contact Dermatitis 2004; 50: 304-213.

Figure 1. Sources of variability in atopy patch testing.

Oranje et al. [157] presented the results of their study conducted on 175 patients with AD. They studied the clinical use and reproducibility of SAFT with CM, HE and peanut. They also evaluated the results of SAFT comparing three kinds of application system: original SAFT patches, square chambers [222], silver patches and big Finn Chambers. The concordance among the studies application system was high and the agreement rate was from 92% to 100%. Studying the over-time reproducibility of SAFT with CM (second testing after 1 year) it was observed very high inter-test agreement of 93% (Cohen's $\kappa = 0.87$). The inter-test agreement of SAFT with HE was 94% (Cohen's $\kappa = 0.80$) and using peanut as allergen, this agreement was 88% (Cohen's $\kappa = 0.76$). Authors concluded that SAFT is a suitable test in small children with AD, because it is sensitive and child-friendly, and urticarial flare-ups are common in children suffering from AD [157]. De Ward-van der Spek et al. [51] found good correlation between the results of SAFT, SPT and OFC. According to the achieved results of sensitivity (SE = 83%), specificity (SP = 100%), positive predictive value (PPV =

100%) and negative predictive value (NPV = 91%) the test seemed to be reliable skin test for diagnosing FA in younger children with AD. According to their conclusion, SAFT should be performed in children under 3 years, whereas SPT should be used above this age. However, the results of this study were not confirmed in the following studies of other groups and the use of SAFT was abandoned, since there was a high risk of false-negative results and poor reproducibility [79]. In study of Hansen et al. [79], SAFT did not improve diagnostic capacity of SPT to detect FA also in the younger children and this test showed poor reproducibility. They also observed severe systemic reaction due to SAFT and it makes this test under this condition questionable, especially in small children. Reproducibility of SAFT was insufficient as discordant reactions were observed in nearly half of the patients. Changing the cut-off point (considering isolated erythema as positive result) did not change the outcome of the test. Since with SPT performed with the same testing material they observed excellent result, it is unlikely that false-negative reactions in SAFT were due to the test material, although the use of non-standardized extracts always carries a risk of variability [79].

According to the Position paper of European Academy of Allergy and Clinical Immunology (EAACI) on Atopy patch test technique and also according to the results of some studies, it is recommended to check the application site of APT after 20-30 minutes for possible presence of immediate-type skin reaction, what can be considered as an alternative of SAFT [105, 143, 220].

5. Skin Immunology of Atopy Patch Test Reactions

The APT reaction is based on cutaneous T-cell mediated responses after epicutaneous application of an intact protein allergen [241]. In atopy patch testing, the allergens are whole allergens applied on the unaffected and non-abraded skin under occlusion. The allergen consequently penetrates the epidermis, where it is thought to be captured by IgE molecules which then bind to high-affinity IgE receptors on the Langerhans cells [85, 150]. Allergen-specific T-cells are thereby activated and initiate an eczematous reaction, which immunohistochemically is nearly the same as that found in atopic dermatitis lesions [217]. It was demonstrated many times, that the APT reactions are associated with the T lymphocyte-mediated allergen-specific response [241]. The close macroscopic and microscopic similarities between the specimens from APT sites and lesional skin of patients with AD indicate that APT is a valid model to study allergic inflammation in AD [112]. Clinical investigations showed that positive APTs (with T-cells infiltration in the skin) correlate with clinical late responses [140].

Biopsy specimens of the APT test sites in patients with AD were found to have initial T_H2 cell infiltration, followed by a predominance of T_H1 cytokines and eosinophils [40]. Similar biopsy findings have been observed in the skin of AD patients during acute and chronic lesions. T_H2 cytokines predominate earlier in the disease, with a transformation to T_H1 predominance in the chronic lesions [117]. In the freshly induced APT lesions, a Birbeck granule negative, non-Langerhans cell population with an even higher IgE-receptor expression that the Langerhans cell, the so-called inflammatory dendritic epidermal cells (IDEC) has recently been demonstrated [104, 243]. This phenomenon occurred both in non-atopic and atopic AD [104]. This may also explain IgE-associated activation of allergen-specific T-cells finally leading to eczematous skin lesions in the APT [181, 226]. Immunohistolochemical studies of biopsied positive APT reactions demonstrated a mononuclear cell-infiltrate in the upper part of dermis, consisting mainly of T-cells, with a slight predominance of T-helper-cells as compared to T-suppressor cells, and about 10% $CD1^+$ cells. No significant responses were obtained in peripheral blood mononuclear cell-cultures stimulated with the various allergens [110]. Negative APT sites were immunohistochemically similar to clinically noninvolved AD skin. There were no significant differences between patients with AD who had positive and negative APT regarding either clinical features, the composition of cellular infiltrate, or the presence of allergen-specific T-cells in clinically non-involved skin. However, differences were observed regarding the presence of IgE on epidermal $CD1^+$ cells. These results indicates that a positive APT reaction requires the presence of epidermal IgE^+CD1a^+ cells in clinically noninvolved skin, but also that other, as yet unknown, discriminatory factors are involved [113]. In eczematous reactions following aeroallergen application, antigen-bearing Langerhans cells, co-expressing IgE, were exclusively found in the epidermis after 6-h, and mainly in the dermis after 24- and 48-h [70, 213]. Acute skin lesions in AD are characterized by $CD4^+$ T_H2 lymphocytes and eosinophils and the release of T_H2-type cytokine (e.g. IL-4, IL-5, and IL-13) [55, 72, 217]. Chronic lesions show a predominance of T_H1-lymphocytes, macrophages, and T_H1-type cytokines (e.g. IL-2, IL-12, IFN-γ) [126]. The biopsies taken at 24-h after the patch test reaction to aeroallergens showed spongiosis with intra-epidermal vesicles. The cellular infiltrate in these vesicles consisted mainly of eosinophils, mononuclear cells (especially T-lymphocytes) and IgE-bearing Langerhans cells. After 48-h any eosinophils were present. In the dermis also a strong eosinophil infiltrate was observed, which declined after 48 h [18]. In APT reactions sites within 72-h the migration of inflammatory dendritic epidermal cells occurs as an early event. The specific up-regulation of FcεRI, especially on the IDECs, occurs later during formation of extrinsic but not intrinsic AD lesions induced by APT. A late event, the alteration of the dendritic cells phenotype with increased expression of CD36 occurs [104]. Numerous lymphocytes migrated into positive reactions sites and nearly all belonged to the T-cells. Some eosinophils increase was also apparent. Immunoglobulins (IgG, IgM, IgE) were present in normal amounts in untouched sites, but many times more of all three types were present in positive patch test sites. Also increased amounts of complement C3 and C4 were released into positive patch test sites. In APT, all the four type of allergic reactions according to Gell and Coombs are involved and are important in provoking visual skin reaction [16]. The APT reactions showed great similarity to lesional AD skin macroscopically showing erythema, induration and papules. Microscopically the dermal infiltrate consisted of T-cells, eosinophils but with hardly any neutrophils, resembling lesional skin [153].

The results of the study of Thepen et al. [217] showed that in lesional skin IFN-γ-positive cells predominated over $IL\text{-}4^+$ cells. The $IFN\text{-}\gamma^+$ cells are mainly $CD\text{-}3^+$ and $CD\text{-}4^+$, and the rest were $CD\text{-}8^+$, $RFD\text{-}1^+$ and $RFD\text{-}7^+$ cells. The $IL\text{-}4^+$ cells are exclusively $CD\text{-}4^+$ cells; no eosinophils or mast cells were found to stain for IL-4. With regard to the T_H-cell response, a clear dichotomy of the eczematous response to the aeroallergens in the skin was observed. In the initiation phase of IL-4 production by T_H2 and T_H0 cells is predominant over IFN-γ by T_H1 and T_H0 cells. In the late and chronic phases, the situation was reversed and IFN-γ production by T_H1 and T_H0 cells predominated over IL-4 production by T_H0 and T_H2 cells. The kinetic of this increase in inflammatory cells was clearly demonstrated during the APT. T-cells numbers increased rapidly until they reached a maximum at 24-h for $CD\text{-}8^+$ and at 48-

h for CD-4^{+} cells. Eosinophil numbers peaked at 24-h APT after which they decreased rapidly. A continuing increase in IFN-γ and IL-4 positive cells was observed, although the increase in IFN-γ cells was larger than that of IL-4 between 24-h and 48-h APT [217]. It was shown a relationship between circulating, specific T-cells with T_H2-like cytokine response and positive APT reactions [97]. However, there are dramatic differences in skin T-cells that are cutaneous leukocyte antigen (CLA) positive [162] and gastrointestinal T-cells, which are primarily gamma-delta T-cells with a higher proportion of CD45RO positive cells. In addition, gastrointestinal T-cells have a different homing adhesion molecule [50, 76, 100]. These observations make it difficult to rationalize how T-cells reactions in the gastrointestinal tract could be detected in the skin. One possible explanation is the migration of T-cells throughout the circulatory system or the interaction of T cells in regional lymph nodes [60].

Grewe et al. [72] analyzed the cytokine pattern expressed *in situ* in APT with aeroallergens. In 24-h house dust mite APT reactions, expression of IL-4 mRNA and IL-2 mRNA increased, but IFN-γ did not, as compared with control skin. But in 48-h APT reactions, IFN-γ mRNA and IL-2 mRNA expression were increased above the levels observed in control skin, whereas IL-4 mRNA expression was decreased bellow background levels. These data demonstrated that a switch from a T_H2-like to T_H1-like cytokine response occurred in aeroallergen APT in AD patients. This biphasic pattern was specific to inhalant allergen APT reactions, as it was not observed in irritant reactions in the same patient. IFN-γ production by T-cells may be induced by the cytokine IL-12. Up-regulation of IFN-γ mRNA expression in APT reaction sites was preceded by an increased expression of p35 subunit of IL-12 mRNA. These observations suggest that increased IL-12 expression may contribute to the observed switch of the *in situ* cytokine secretion pattern.

Buckley et al. [24] compared the immunohistological differences between APT positive sites, negative sites and uninvolved skin in AD patients after the application of aeroallergens. All positive patch tests exhibited characteristics of a cell-mediated immune response. The negative patch test sites were also found to contain evidence of mononuclear cell infiltration. Both negative and positive patch test sites showed significantly greater proportion of T-cells compared to uninvolved skin. No increase in numbers of RDF1^{+} and RDF7^{+} macrophages were observed in either positive or negative patch test sites. Expression of CD23 by CD1^{+} Langerhans cells was raised in both negative and positive patch tests compared to uninvolved areas. A significant increase in RDF7^{+} and CD23^{+}cellular population was seen in positive patch test sites compared to uninvolved skin. An increase in the proportion of RDF1^{+} cells expressing CD23 was seen in both negative and positive patch tests compared to uninvolved skin. The presence of systemic cell-mediated immunity to specific allergens identified in patients by positive patch test may also be present when no clinical signs are seen at patch test sites.

Holm et al. [85] studied the changes in the cellular infiltrate in APT in two small groups of patients with or without the detectable serum IgE to *Dermatophagoides pteronyssinus*. The number of IgE^{+}, CD-4^{+}, Metalloproteinase-9^{+} cells and eosinophils increased in the subgroup with detectable specific IgE. FcεRI^{+} and CD-8^{+} cells increased with time in both groups. A correlation was found between the levels of specific IgE and the score of dermal cell infiltrates at 72-h. Also this study strengthened the hypothesis, that the IgE molecule has a key role, at least as an amplifier, in the APT reaction. However, the presence of allergen-specific IgE does not seem to be obligatory for the APT reaction, as it takes place in individuals without detectable levels of allergen-specific IgE in serum [34, 88, 129, 192] and people with elevated levels of specific IgE do not necessarily show a positive APT reaction [98]. T_H2-like response of peripheral blood mononuclear cells to some aeroallergens showed stronger correlation to the APT reaction than it did to the specific IgE-levels or SPT reaction [97]. This indicates that the T-cell response is another crucial factor for allergen-specific eczema reactions [85]. Positivity of APT was not associated with IFN-γ or IL-4 release from peripheral blood mononuclear cells with or without stimulation with CM antigens [106].

Animal model of APT with aeroallergens has shown histologically superficial perivascular mononuclear and granulocytic dermatitis development after 6-h, and progression in severity at 24-h. Additionally, at 48-h epidermal spongiosis, hyperplasia, and pustules were developed. At and beyond 6-h, progressive $CD1^+$ epidermal Langerhans cells hyperplasia with cluster formation and dermal dendritic cell infiltration was observed. mRNA expression for a cytokines IFN-γ, IL-6, IL-12p35, IL-13, IL-18, and thymus activation regulated chemokine (TARC) exhibited significant increases during the challenge compared to baseline, but there was no appreciable alteration in expression for tumor necrosis factor-alpha (TNF-α), IL12p40, IL-10, regulated on activation normal T-cell expressed and secreted (RANTES), IL-5, IL-2, IL-4, and IL-8. No correlation was found between clinical scores of positive reaction and cytokines. These authors concluded that IL-6 plays a role in early reactions followed by an increase of TARC and IL-13, while IL-18 progressively increases in later reactions [126]. In another animal model, macroscopic and microscopic positive APT reactions occurred only whenever serum positive IgE was present against the tested allergens. Microscopically, positive APT was associated with epidermal hyperplasia, Langerhans cell hyperplasia, and eosinophil and lymphocyte epidermotropism. Dermal inflammation was mixed and arranged in a superficial perivascular to interstitial pattern. Numerous IgE^+-$CD1^+$ dendritic cells and gamma-delta T-lymphocytes were observed. Macroscopically and microscopically, APT reactions in these experimentally sensitized animals resembled those seen in lesional biopsy specimens of dogs and human with spontaneous AD. The APT provides a relevant experimental model to investigate the pathogenesis and treatment of both canine and human AD skin lesions [156].

Gfesser et al. [65] compared the difference in the disturbance of skin barrier in atopy patch test reactions with classical patch test with chemical substances. The barrier disturbance was evaluated by measurement of transepidermal water losses (TEWL). The TEWL of the positive APT was significantly higher in comparison with the control sites or with positive patch test reactions with chemical contact allergens. In AD, the epidermal barrier function is altered only in the positive APT reactions, in contrast to positive patch test reactions to contact allergens. As a consequence of this aeroallergen-induced altered epidermal barrier function, further allergens can more easily penetrate the skin, including a vicious circle and perpetuating the eczematous reactions.

The macroscopic and microscopic resemblance of the patch test reaction to aeroallergens and food allergens in clinically non-involved AD skin with lesional AD skin suggests that the patch test reaction to aeroallergens and food allergens may be a helpful *in vivo* model to further study the pathogenesis of AD [18, 19, 30].

6. Technical Aspects of Atopy Patch Testing

There are many technical aspects of atopy patch testing which might have important impact on the final outcome of this testing procedure, what was demonstrated in many studies. It is very important to take into consideration all of these factors and attributes, as only by this, it is possible to find the right place of the APT in the diagnostic algorithms of particular disease. There are many important factors influencing the final results of APT and the most relevant are application site, chamber size and type of application system, source, concentration and preparation of the allergens, vehicle used, negative control, pretreatment of the testing site, occlusion time of application and the time of reading of induced eczematous reaction, applied scoring system, medicament withdrawal and many other aspects and factors.

A. Application Site

Although most investigators performed the APT on the upper back of the patients [104, 111, 123, 124, 192, 230], some authors studied as well other possible locations, such as antecubital [111, 149] or popliteal fossa [149], outer upper arm [16], or the extensor side of the forearm [70, 81, 236]. Langeveld-Wildschut et al. [111] tested the uninvolved skin of the back and the antecubital fossa in 10 AD patients, failing to demonstrate the differences in response between these two test sites. Norris et al. [149] described an increased incidence of immediate pruritic reaction on the antecubital fossa compared with the back after epicutaneous allergens′ application to the uninvolved skin, probably due to local differences in cutaneous absorption [57], itch points [178] or mediators of pruritus [48]. Although the back seems the most practical location for testing, it is also suggested that the best reproduction of AD requires various conditions, such as site of normal distribution of the lesions [70]. Weissenbacher et al. [236] in their study compared different application sites. There was no significant difference in either the intensity or quality of itch between the two forearms and the back. Regarding the quantification of the test reaction, a positive reaction was more frequent on the back (94% vs. 69% on the forearms) and the peak atopy patch score was higher on the back compared with both forearms. Allergens should be therefore preferably applied on the back for the APT.

Van Strident and Korstanje [227] investigated the existence of intra-regional variations in response to allergens applied on the back. The results were assessed by classical visual scoring system and the skin blood flow at the test sites was quantified by laser Doppler flowmetry. The results of laser Doppler flowmetry showed highly significant differences between the upper and lower back, however, the differences in the visual scores were not demonstrated. It was confirmed that the upper back seems to be one of the most sensitive areas [119, 120]. It is necessary to study also the differences between left and right side of the back [227]. Apart from regional differences, intra-regional variations in the response both to the allergens and irritants have been described on forearm skin [115, 193, 224].

The most advantageous place might be the patient's upper-back with uninvolved skin without AD lesions [105, 143, 219, 220].

B. Chamber Size and Atopy Patch Test Application System

The chamber size (diameter) is another very important parameter influencing the outcome of the APT. The studies differ in the diameter of used chambers or in the APT application system. Many authors used and recommended the 12-mm aluminium cups Finn Chambers (Epitest Ltd., Hyrylä, Finland; Hermal, Reinbeck, Germany; Finn Chambers, Allerderm laboratories, Sonoma, Calif, etc.) on Scanpor tape [45, 60, 79, 92, 93, 123, 172, 174], but the other diameters of the testing chambers have been studied as well: 8-mm aluminium cups [32, 106, 158], 9-mm square IQ Chambers [197], 10-mm plastic quadratic cups [95, 175, 176], 11-mm Finn chambers fixed with stretch plaster on the patients′ back [104], Finn Chambers of Chemotechnique Diagnostics [108] or Curatest adhesive strips [95, 169].

Since the small backs of young children offer little space for atopy patch testing, Niggemann et al. [144] investigated and compared 6-mm chambers with classical 12-mm aluminium cups traditionally used in the previous studies. Sensitivity, specificity, negative and positive predictive values and also efficiency results showed that the 12-mm chamber size yielded much better results than the 6-mm chamber size. The use of 6 mm aluminium cups may result in a decreased sensitivity compared to the use of 12-mm cups, which should be therefore preferred [141]. The 8-mm aluminium cups are recommended for children younger than 3 years and standard 12-mm aluminium cups for children older than 3 years [32, 197]. However, it was clearly showed, that the diagnostic accuracy of the APT using a 12-mm Finn chambers was greater than for the 6-mm chamber [144].

Just a few data exist on the use of alternative materials such as quadratic cups [95, 175, 176] now available on the market [141].

Our Data

We conducted two studies on large unselected populations of school children aimed on clinical characteristics of atopy patch tests with food and inhalant allergens. Our first study comprised all four-grader children attending schools in four small semi-rural towns in northern Slovakia (Poprad, Dolny Smokovec), and central Italy (Ronciglione, Caprarola). This study was conducted between October 2002 and February 2003. The total number of examined children was 335 (46% boys, age 10.10±0.63 years): 185 from Slovakia (50.3% boys, age 10.11±0.68 years) and 150 from Italy (40.7% boys, 10.09±0.55 years) [174]. Our second study comprised 532 children (50.6% boys, age 10.23±2.27 years) attending three grammar schools in the north of Rome (Italy) from October 2005 to March 2006 [94, 95, 175, 176]. The atopy patch tests and skin prick tests with food and inhalant allergens were performed according to the recommendation of EAACI. The studies were approved by the Ethical Committee of the Pediatric Clinic of Rome University "La Sapienza".

In our study we applied fresh food allergens on both sides of the back with two kinds of APT devices: plastic quadratic cups on hypoallergenic textile tape of 10-mm diameter (Finn Chambers, Haye's, Alphen, the Netherlands) in 293 children (age 8.71±1.74 years, 54.1% boys) or with fine blotting paper circles on transparent adhesive tape (Curatest, Lohman & Rauscher SRL, Padova, Italy) in 66 children (age 8.16±1.74 years, 47.2% boys). Using Finn Chambers, we observed 8.9% positive APT results for CM and 6.1% for HE. 4% children had positive APT with tomato and 6.1% with wheat flour. In contrast, when we applied the allergens with Curatest, the frequencies of positive APT were: 12.1% for CM, 12.1% for HE,

19.7% for tomato and 13.6% for wheat flour (table 1). We did not observe the differences in the frequencies of positive APT results between left and right side of the back. In general, we obtained more positive results with Curatest than with Finn Chambers for all four food allergens. This difference reached statistical significance only for tomato ($p < 0.001$), whereas for other three allergens these differences were not statistically significant (figure 2)

Table 1. Frequency of positive APT results with fresh food allergens applied by two different devices.

	Finn Chambers®		Curatest®	
	Left side	***Right side***	***Left side***	***Right side***
Cow's milk				
① neg.	249 (85.0%)	255 (87.0%)	52 (78.8%)	55 (83.3%)
② ?	18 (6.1%)	12 (4.1%)	6 (9.1%)	4 (6.1%)
③ +	12 (4.1%)	14 (4.8%)	0	0
④ ++	11 (3.8%)	10 (3.4%)	7 (10.0%)	5 (7.6%)
⑤ +++	3 (1.0%)	2 (0.7%)	1 (1.5%)	2 (3.0%)
Pos. (③+④+⑤)	26/293 (8.9%)	26/293 (8.9%)	8/66 (12.1%)	7/66 (10.6%)
Hen's egg				
① neg.	256 (87.4%)	264 (90.1%)	54 (81.8%)	57 (86.4%)
② ?	19 (6.5%)	14 (4.8%)	4 (6.1%)	5 (6.1%)
③ +	15 (3.1%)	8 (2.7%)	0	0
④ ++	6 (2.0%)	6 (2.0%)	6 (9.1%)	3 (4.5%)
⑤ +++	3 (1.0%)	1 (0.3%)	2 (3.0%)	2 (3.0%)
Pos. (③+④+⑤)	18/293 (6.1%)	15/293 (5.1%)	8/66 (12.1%)	5/66 (7.6%)
Tomato				
① neg.	236 (93.3%)	235 (92.9%)	50 (75.8%)	50 (75.8%)
② ?	7 (2.8%)	8 (3.2%)	3 (4.5%)	1 (1.5%)
③ +	4 (1.6%)	3 (1.2%)	0	1 (1.5%)
④ ++	4 (1.6%)	4 (1.6%)	10 (15.2%)	12 (18.2%)
⑤ +++	2 (0.8%)	3 (1.2%)	3 (4.5%)	2 (3.0%)
Pos. (③+④+⑤)	10/253* (4.0%)	10/253* (4.0%)	13/66 (19.7%)	15/66 (22.7%)
Wheat flour				
① neg.	262 (89.4%)	262 (89.4%)	53 (80.3%)	56 (84.8%)
② ?	13 (4.4%)	13 (4.4%)	4 (6.1%)	4 (6.1%)
③ +	7 (2.4%)	7 (2.4%)	0	1 (1.5%)
④ ++	8 (2.7%)	9 (3.1%)	4 (6.1%)	3 (4.5%)
⑤ +++	3 (1.0%)	2 (0.7%)	5 (7.6%)	2 (3.0%)
Pos. (③+④+⑤)	18/293 (6.1%)	18/293 (6.1%)	9/66 (13.6%)	6/66 (9.1%)

* For technical reason, we did not perform APT with tomato in 40 children.

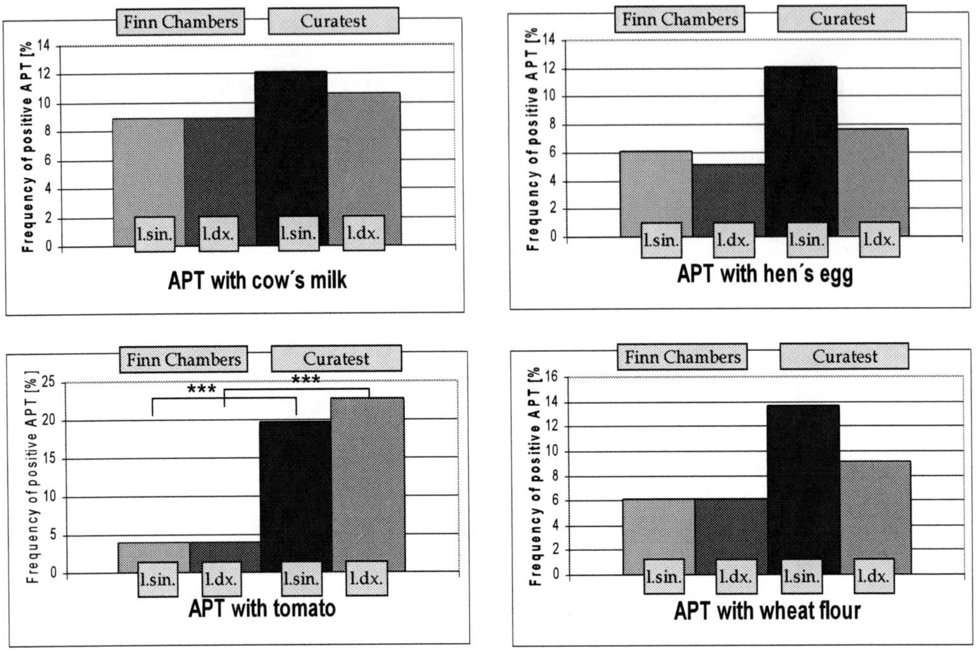

*** p < 0,001

Figure 2. Frequencies of positive APT results with fresh food allergens applied by two different devices on the left and right side of the back.

We studied the influence of used APT device on the right-versus-left reproducibility. In the first group we applied on both sides of the back device Finn Chambers (N = 205, 54.1% boys, 8.71±1.43 years), in the second group we applied the allergens with device Curatest (N = 70, 47.2% boys, age 8.16±1.74 years) and finally in the third group of children we used the combination of both APT devices (N = 158, 47.5% boys, age 11.43±1.43 years).

The average value of Cohen's κ in the first group with two Finn Chambers on the back was 0.616 (0.447 - 0.810) depending on the allergen used. The highest value was reached with CM (0.810, highly satisfactory agreement) and the lowest with wheat flour (0.447, fairly satisfactory agreement). The percentage of reproducible positive APT results was from 30.8% (tomato) to 70.8% (CM), what means that from 29.2% to 69.2% of positive results had negative APT results on the opposite side of the back (table 2).

In the second group with two Curatest devices, the average Cohen's κ value was 0.355 (0.323 – 0.406). The highest value was reached for this device with HE (0.406, fairly satisfactory agreement) and the lowest with CM and wheat flour (0.323 and 0.327, barely satisfactory agreement). Approximately 70% of positive results were non-reproducible (tab. 2).

Analyzing the reproducibility in the group with combined APT devices, we obtained the lowest average Cohen's κ value of 0.324 (0.162 – 0.553). The highest value was reached with tomato (0.553, fairly satisfactory agreement, 88.9% of reproducible positive APT results) and the lowest with CM (0.162, unsatisfactory agreement, only 10% of reproducible results). From this point of view, the use of Finn Chambers is the most advantageous in comparison with Curatest (table 2).

Table 2. Reproducibility of atopy patch tests with fresh food allergens according to the APT device used

	2 x FinnChambers®				**2 x Curatest®**				**Combination of both APT devices**			
	P/P	N/N	P/N N/P	Cohen's κ	P/P	N/N	P/N N/P	Cohen's κ	P/P	N/N	P/N N/P	Cohen's κ
CM	17/196 (8.7%)	172 (87.8%)	7 (3.5%)	**0.810**	3/66 (12.1%)	54 (81.8%)	9 (13.7%)	**0.323**	2/152 (1.3%)	135 (88.8%)	15 (9.9%)	**0.162**
HE	9/196 (4.6%)	178 (90.8%)	9 (4.6%)	**0.642**	3/66 (4.5%)	56 (84.8%)	7 (10.6%)	**0.406**	8/152 (5.3%)	116 (76.3%)	28 (18.4%)	**0.256**
TO	4/196 (2.0%)	183 (93.4%)	9 (4.6%)	**0.447**	7/66 (10.6%)	45 (68.2%)	14 (21.2%)	**0.366**	8/152 (5.3%)	133 (87.5%)	1 (7.2%)	**0.553**
WF	8/196 (4.1%)	177 (90.3%)	11 (5.7%)	**0.553**	3/66 (4.5%)	54 (81.8%)	9 (13.6%)	**0.327**	7/152 (4.6%)	124 (81.6%)	21 (13.8%)	**0.325**

CM cow's milk, HE hen's egg, TO tomato, WF wheat flour, P positive, N. negative

C. Ready-to-Use Atopy Patch Tests

Atopy patch test has an advantage that it is non-invasive and is used on the skin. This is a great advantage especially in children. Indeed, the skin is at one and the same time a barrier of the organism with a local immune system and a possible route of administration via transdermal passage of drugs [9]. So allergens put on the skin can come into contact with immune cells, however, at the same time, their passage through the skin can be limited by the very low permeability of the skin [202]. Classical occlusive delivery system for APT – aluminium cups Finn Chambers placed on hypoallergenic tape is intended to be prepared and secured on the skin of children by the medical practitioner. These tests, despite of many efforts to complete the standardization [34, 81, 105, 220], still suffer from many drawbacks. To solve these problems with standardization, some innovative ready-to-use APT has been developed, e.g. the E-patch® system or Diallertest®. These APT devices are composed of an occlusive chamber and differ from those previously described in that it is ready-to-use, with the allergen present as dry milk powder, with defined protein content and a guaranteed protein quality. Because of skin occlusion and normal skin transepidermal water loss (TEWL) [1, 7], this system is intended to progressively create a solution of the proteins contained in the milk powder at the surface of the skin. Ideally, the allergens should concentrate in the epidermis to trigger a local reaction but nor permeate the skin, to avoid any possible systemic anaphylactic response [202]. This test has great accuracy because it maintains protein allergens in their native state without using preservatives or excipients. It was designed to be easier to use than the atopy patch tests usually used to detect pediatric cow's milk allergy.

In the study of Kalach et al. [101], a ready-to-use atopy patch test - Diallertest® (DBV-Technologies, Boulogne-Billancourt, France) was compared with another APT device, the classical Finn Chambers, in pediatric cow's milk allergy. This study involved 49 children with confirmed CMA manifested by atopic dermatitis, digestive manifestations, or both. All the children underwent both APT techniques, with a reading at 72 hours after application and followed by a milk elimination diet and open cow's milk challenge. A positive results were seen in 44.8% of children with the ready-to-use set and in 26.5% of children with comparator classical APT. Both techniques were concordant in 67.3% patients. Sensitivity was much better with ready-to-use set than with standard APT (76% vs. 44%, $p = 0.02$), whereas specificity was 93.8% for botch techniques. With ready-to-use APT they observed 1 false-positive and 6 false-negative results, whereas with standard APT they found 1 false-positive and 14 false-negative results. Test accuracy was also better for ready-to-use APT (82.9% vs. 63.4%, $p = 0.05$). The ready-to-use set exhibit better sensitivity with no side effects and seems to be more practical than classic APT. This ready-to-use CM APT called Diallertest is 26-mm in diameter and consists of three parts: 1) a central transparent membrane of polyethylene charged with electrostatic forces able to retain a powdered material for a long period and to allow a direct visual monitoring of cutaneous reaction, 2) a biadhessive mousse layer enclosed by 2 liner sheets, and 3) an adhesive sheath of non-woven film. The same mixture used for classical comparator APT was deposited on the central plastic support in the form of microgranules forming a homogenous monolayer retained by electrostatic forces. Each APT thus contained 250 μg of CM proteins with 60% casein and 40% lactoserum proteins. The ready-to-use APT serving as negative control had the same structure but was deprived of any CM powder. This study introduced a ready-to-use novel type of patch testing. The basis of the test relies on the ability to provide to the skin intact molecules devoid of any solvent and able to be solubilized through the sole sweat secretion. This technique exhibits several advantages compared with the classical APT, both in terms of practical easiness and of standardization. The amount of milk deposited on the patch is constant and easily measurable. When applied, the device delivers to the skin the total amount of milk deposited on the patch, whereas in any other kind of patch testing, the exact amount of food allergen delivered to the skin is difficult to assess: when present in the form of pureed food, only a part of the food comes in close contact with the skin, and when the food is deposited on a blotting paper, a large amount remains inside the paper. The standardization of APT technique requires not only standardizing the amount of antigen deposited in the device but also the amount of antigen able to reach immunocompetent reactive cells. The advantage of this new APT device is that it could be applied by the parents to the child 3 days before the reading by the physician, thus avoiding second consultation. The transparent plastic membrane also allows direct visual effects monitoring of immediate or delayed cutaneous reactions without the removal of the set. This test might be proposed as a first-line diagnosis [101]. It means that Diallertest® in comparison with classic APT is not only more convenient but also more accurate [136].

In another study conducted by Soury et al. [202], the localization of β-lactoglobulin delivered into the skin by an innovative ready-to-use APT E-patch® was investigated. As this test allows the detection of delayed allergies at the skin level, they also studied the efficacy and safety of this new device. In this study, E-patch® containing β-lactoglobulin was placed for 24 hours in contact with hairless rat skin. Transdermal passage was monitored by measurement of β-lactoglobulin A-[methyl-^{14}C] or by two-site enzyme immunoassay. After 24 hours, 92% of β-lactoglobulin remained on the skin. The iterative skin stripping showed a 135-fold higher concentration of β-lactoglobulin in the stratum corneum than that found in the epidermis-dermis. Analysis of the solution in the receiver compartment by radioactivity assays or immunoassays indicates that intact protein did not cross the skin. Authors concluded that new ready-to-use E-patch® system allows native β-lactoglobulin to concentrate in the stratum corneum, in the vicinity of immunocompetent cells, but does not lead to its systemic delivery. Therefore, it was suggested that this delivery system creates ideal conditions for promoting a positive topical response with reduced risk of systemic anaphylactic reactions caused by the native form of the β lactoglobulin.

D. Marking of Atopy Patch Test Testing Sites

Important technical part of performing the APT is marking the application sites to indicate where the different allergens have been tested. Marking the skin is necessary when several readings are carried out [116, 104]. Very useful is the reading with provided Finn Chambers reading plate, which helps locate test sites and identify the reactions.

E. Allergens Tested in Atopy Patch Test

Primarily, the APTs were extensively studied with aeroallergens, which are also very important in the pathogenesis of AD. Later on, also the APTs with food allergens have been introduced into clinical use. Many food allergens have been applied such as cow's milk, hen's egg (yolk, white), soy milk, peanuts and other nuts, cereals (wheat, rice, maize, buckwheat, barley, oat, rye), vegetables (tomato, potato, soybean, cabbage, carrot), fruits (apple, orange, banana), meat (pork, beef, cod, chicken), cacao, mustard powder. In some studies, high number of food allergens (25-50) has been investigated simultaneously [16, 173]. Any food can be assessed with APT, although cow's milk, hen's egg, wheat, and soy have been studied most extensively. It seems that the number of positive APTs is dependent from the type of used allergen [48]. Only those allergens known to precipitate the dermatitis or other symptoms by history or which were identified in patients′ home environments elicited positive patch reaction. There is no reason to doubt that many other allergens work in APT testing [60, 141, 206].

F. Allergens Preparation and Their Concentration

Technical aspects of APT with regard to vehicle, dose response, mode of application, reading times, scoring and clinical covariates have been studied [34, 35, 38, 84, 153]. Further standardization of the test procedure and evaluation of the APT reactions are needed, especially with regard to the choice of allergen extracts. This is the crucial aspect of APT with food allergens, which requires further investigations, since optimal testing substances with food allergens are not yet available. According to the Position paper of EAACI [220], the use of standardized preparations in petrolatum is recommended. However, the availability of these testing substances is poor, and according to the several studies, their use in comparison with foods in their native forms does not increase the diagnostic accuracy of the APT. There are still some methodological problems in performing APT. Commercial preparations for food APT are poorly available and preparations for many food allergens are still lacking. Allergen concentration in preparation for SPT is considered too low and not suitable for APT. The use of fresh food allergens is recommended, but possible variations in allergen content according to the diet of the animals should be taken into account [67].

In the literature, many modifications of the allergens preparation could be found: 200 mg of cereals as flour mixed in 0.2 mL of isotonic saline solution [93], 300 mg of wheat in 0.2 mL isotonic saline [103], 300 mg of skimmed milk powder in 0.2 mL of isotonic saline solution [32, 93, 103, 123, 211], 40 mg of egg powder in 0.2 mL isotonic saline [103, 123, 211], 200 mg of egg white, cereals in 0.2 mL of isotonic saline to make a "porridge" that remains on the test cup [79, 92, 106], 20 μg of cow's milk powder containing 35 g of protein/100 g mixed with saline [230], wheat gluten flour suspension (1 g in 10 ml of normal saline) [8, 80, 130, 158, 172], 3 g of powdered skim milk mixed with 1 mL of isotonic saline solution [203, 205, 206], egg white, celery and wheat flour, all at 1/3 w/v in petrolatum [104], 2 g of dry food (soy infant formula and skim milk powder, dried egg white, wheat, oat, barley, rye, and rice flour, corn meal, dehydrated potatoes) in 2 mL isotonic saline solution [204], a mixture with egg yolk or white with petrolatum oil 2:1 [67], CM preparation in petrolatum (30%) [103], CM, HE and wheat flour preparations (10%) and tomato preparation (20%) in petrolatum [175, 176], the commercial extracts to fish or cooked suspected fish (ev. boiled) [169], hen's egg white in a cream in concentrations up to 1000 times higher than the concentration used for SPT [110], a mixture consisting of two thirds of a powdered CM product and one third of a hypoallergic infantile CM formula diluted in water (13.5 g/100 mL) [101], a mixture of raw peanuts and petrolatum 2:1 [195], allergen preparation in petrolatum and petrolatum oil [125], or non-fat milk in saline and in petroleum [82].

Breneman [16] developed a special kind of food patch test named dimethyl sulfoxide food test (DMSOFT). Fifty sterile, freeze-dried, pure food products were individually suspended in 90% dimethyl sulfoxide (DMSO), which was used to carry both water-soluble and fat-soluble pure food products through the skin to the subcutaneous tissue. Its fat-solvent capabilities and its ability to penetrate intact skin as a carrier vehicle were known. The mixtures were prepared from 1 g of freeze-dried food in 90% DMSO (1 g/5 mL).

Many authors recommend the use of fresh food allergens: fresh pasteurized whole-fat cow's milk containing 3.5% fat, hen's egg (fresh whole whisked raw egg), soy milk [8, 17, 25, 79, 80, 130, 158, 166, 201], raw peanut powder, wheat powder, mustard powder without dilution in saline solution [173], or single-ingredient baby food-fruits, vegetables and meats were placed into the chambers undiluted [204]. The good percentages of sensitivity and specificity for the APT may be explained by the fact that the identical native food allergens are used also for DBPCFC and SPT, so the results of all the three tests are better comparable. Native food seems therefore be most suitable for performing APT [141]. Therefore, it would be necessary to investigate the optimal allergen concentration in commercially available preparation, for example in petrolatum.

The positive APT response rate seems to be strictly dependent on the patch test material. Therefore, the choice of allergenic extracts still remains upon a question. Manzini et al. [125] observed different frequencies of positive APT using the same allergens from three various commercial sources. Standardization of food APT seems to be very difficult. Cereals seem to give positive reactions too easily (probably also due to non-specific irritation) and the results need to be checked with DBPCFC [103].

Cow's milk formula powder, bovine serum albumin, crystallized bovine β-lactoglobulin and bovine casein were mixed with 0.9% saline solution in a separate tube in which a filter-paper disc was placed for an hour to become saturated with the test substance [180]. This filter-paper disc was than placed under usual aluminium cups on the patient's back and no differences in the prevalence of positive APT with whole milk or with CM protein fractions.

Several commercial APTs containing freeze-dried food extracts (so-called Ready-to –use APT) are now available, but their diagnostic accuracy is still largely undefined. A recent pilot study demonstrated the safety and efficacy of one of these ready-to-use APTs (Diallertest) [101]. A study of Berni Canani et al. [8] was aimed to examine the diagnostic accuracy of APTs using fresh food vs. commercially available freeze-dried purified food extracts contained in a commercial kit (Atopy line 1, Euromedical, Lecco, Italy). The freeze-dried purified food protein extracts (Allergon AB, Angelhom, Sweden) were provided in a 20% protein concentration in a vaseline mixture. To exclude false positive reactions, they also tested allergens in a 1:10 dilution. The results from this study showed that the diagnostic accuracy of an APT with fresh food is much better than that of a commercial APT assays with freeze-dried purified food extracts. The differences could be due to many factors, including protein purification procedure, antigen concentration, and capability of penetrating the skin. APTs should be standardized not only for the amount of antigen deposited in the Finn Chamber but also for the amount of antigen able to penetrate the skin and reach the reactive cells.

Laboratories still do not provide standardized food allergens preparations – measuring and codifying their biological activity for the use in APT and further studies are needed for developing the production of better test materials [27]. Many foods in APT could be appropriate in concentrations similar to what is ingested, conversely, for grains, egg, and soy the concentration of 1 g/1 mL has been proposed as ideal. Milk requires higher concentration at 3 g/1 mL for ideal testing [204, 205].

There is a large variation in the allergen concentration that is used in the different studies on APT. Some studies used the commercial solution for SPT [133, 149], while the others use 10-1000-fold the concentration of the SPT standardized to one histamine equivalent prick [168], or x100 concentration used for intracutaneous testing [20, 217]. van Voorst Vader et al. [228] compared three concentrations of house-dust-mite allergens (2000, 10 000, and 50 000 AU/ml) using for patch testing and found most positive responses with the highest allergen concentration. Conversely, Langeveld-Wildschut et al. [111] (1995) were not able to confirm an increase positive APT reaction with 100 000 AU/mL instead of 10 000 AU/mL. Darsow et al. [34] performed the APT with different vehicles and allergen concentrations, and found most positive reactions with an allergen concentration of 10 000 PNU/g and petrolatum as vehicle. The concentration of the allergens should be 5 000-7 000 PNU/g or 200 IR/g [41, 43]. Allergen-concentration should be more than 1 000 PNU/g in children and in adults more than 7000 PNU/g. The use of petrolatum as solvent is better than the use of hydrogel [35-39]. For the aeroallergens, in children lower concentrations of allergens in the testing substance are sufficient (1 000-3 000 PNU/g), what is a half of a dose suitable for adults [41]. The outcome of APT strongly depends on the concentration of applied allergen. With 10 000 PNU/g Darsow et al. [34] observed 72% clear-cut positive APT reactions among the patients with AD, whereas with the concentration of 1 000 PNU/g they noticed only 28% positive APT reactions. The concentration of the allergen is important for the outcome of the test. The higher concentration of main allergen is, the higher prevalence of positive skin test results could be observed [27]. High allergen concentrations and 48 h of application increased the number of positive specific APT reactions [228].

There are only few data about the optimal concentration of food used for the APT in children. The optimal concentration of main allergen is necessary to exclude false-positive reactions. It seems that the concentration has influence on the outcome of the APT. Niggemann et al. [140] studied all applied allergens also in a 1:10 dilution in parallel. They observed 23% results in the 10% diluted APT and these were seen in the patients with the strongest reactions. All these patients showed late eczematous reaction in DBPCFC, mostly to the wheat. This indicates that the APT results with undiluted foods are not biased by unspecific, irritative reactions and native food may be used undiluted [140, 141]. Comparing undiluted and diluted food allergens applied in APT setting, there were no great differences in sensitivity, specificity, or positivity rates [79]. According to the results of sensitivity values, for some allergens APT allergen concentration > 200 IR/g may be necessary to demonstrate the sensitization [39, 40, 42]. Further studies with large number of pediatric patients and including food challenges are necessary for food allergens, after better standardization of the allergen content [42].

Our Data

In our study, we applied three kinds of allergens. Food allergens were tested either in their native form: fresh whole-fat cow's milk containing 3.5% fat, whisked raw hen's egg (yolk and white of egg), tomato and wheat flour (dissolved in physiologic saline 1 g/10 ml); or as standardized commercially available preparations in petrolatum: cow's milk 10%, hen's egg 10%, tomato 20% and wheat flour 10% (Lofarma S.p.a., Milano, Italy). In the subgroup of children we used for APT also two standardized inhalant allergen solutions for SPT (*Dermatophagoides pteronyssinus* and mixed grasses: *Avena sativa, Hordeum vulgare, Secale cereale, Triticum sativum*; ALK-ABELLO, Hørsholm, Denmark).

In our unselected population of Italian schoolchildren we observed in 7.4% positive APT results with CM (frequency of strong positive reactions with papules was 4.7%), and in 9.0% positive APT with HE (strong positivity in 7.0% of children). 5.2% children showed positive results with tomato (4.4% strong positive reactions) and in 7.2% we observed positive APT with wheat flour (5.6% strong positive reactions) (figure 3) [175]. These values were observed with fresh food allergens. The frequencies with standardized food allergens were lower for all the four studied allergens but these differences were not statistically significant (data not shown). Two inhalant allergens yielded positive results in 29.8% (*Dermatophagoides pteronyssinus*) a 3.8% (mixed grasses) of studied children.

In our previous study in two European countries, we observed different frequencies of positive APT with fresh food allergens comparing two nationalities. Positive APT reactions with CM were more frequent in Italy than in Slovakia (9.4% vs. 2.2%, p = 0.008) and the same difference was observed also with tomato (9.3% vs. 3.2%, p = 0.035). There was no difference in the prevalence of positive APT with HE (15.3% vs. 12.4%) or with wheat flour (13.3% vs. 8.6%) between two nationalities. In general, Italian children showed more positive APT results with any of the studied food allergens (27.3% vs. 15.7%, p = 0.013) [174].

We studied also the influence of tested allergens on right-versus-left reproducibility. All the 6 allergens (4 food allergens and 2 inhalant allergens) were simultaneously applied on both sides of the back (figure 4). In the first group we tested fresh food allergens (N = 205, 54.1% boys, age 8.71±1.43 years), in the second group we applied standardized food preparations in petrolatum (N = 93, 50.5% boys, age 12.92±0.81 years) and in the subgroup of children we added also two inhalant allergens (N = 131, 58% boys, age 9.40±2.24 years). The allergens were applied with Finn Chambers.

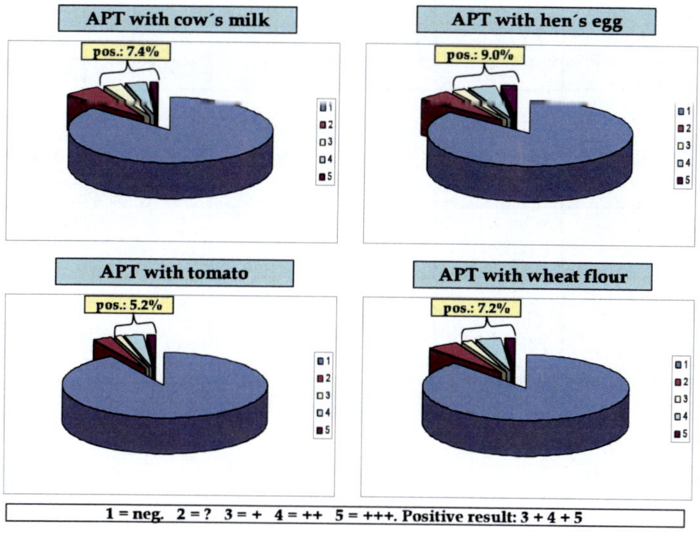

Figure 3. Frequency of positive APT reaction with fresh food allergens in an unselected population of Italian schoolchildren.

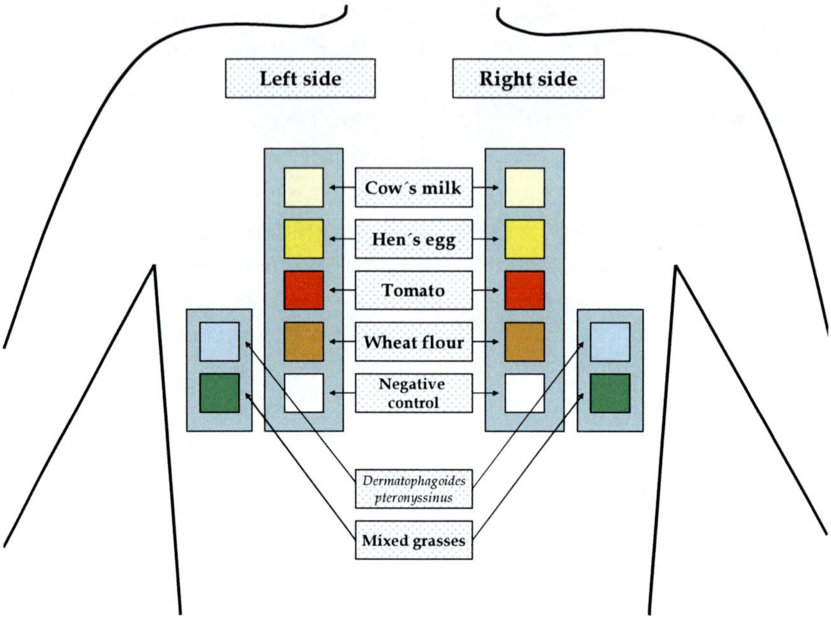

Figure 4. Schema of applied allergens on the back.

Table 3. Reproducibility of food atopy patch tests with native fresh foods and commercially available food allergens preparations in petrolatum

	Fresh food allergens				**Commercial food allergens preparations in petrolatum**			
	Pos./ Pos.	Neg./ Neg.	Pos./Neg. Neg./Pos.	**Cohen's κ**	Pos./ Pos.	Neg./ Neg.	Pos./Neg. Neg./Pos.	**Cohen's κ**
CM	17/196 (8.7%)	172 (87.8%)	7 (3.5%)	**0.810**	4/97 (4.1%)	90 (92.8%)	3 (3.1%)	**0.714**
HE	9/196 (4.6%)	178 (90.8%)	9 (4.6%)	**0.642**	2/97 (2.1%)	93 (95.9%)	2 (2.0%)	**0.656**
TO	4/196 (2.0%)	183 (93.4%)	9 (4.6%)	**0.447**	1/57 (1.8%)	55 (96.5%)	1 (1.8%)	**0.659**
WF	8/196 (4.1%)	177 (90.3%)	11 (5.7%)	**0.563**	3/97 (3.1%)	91 (93.8%)	3 (3.1%)	**0.651**

CM cow's milk, HE hen's egg, TO tomato, WF wheat flour, pos. positive, neg. negative

Using fresh food allergens, the highest Cohen's κ value was reached with CM (0.810, highly satisfactory agreement). For HE Cohen's κ value was 0.642, what indicated satisfactory agreement. Tomato and wheat flour gave fairly satisfactory agreement comparing the APT results on both sides of the back (table 3). The lowest percentage agreement between positive APT results we observed with tomato (only 30.8% positive reproducible results), whereas with CM this agreement was 70.8% (table 2). The reproducibility of positive APT with commercially available standardized food allergens preparations (recommended by EAACI) was lower for all the four food allergens and all Cohen's κ were between 0.651 to 0.714 (satisfactory agreement). In reality, approximately only 50% of positive APT reactions were reproducible (table 3). According to these results, the use of fresh food allergens is more suitable for food atopy patch testing [176].

In the subgroup of children, in which we applied on both sides of the back two inhalant allergens, the concordance between both sides was 100% (Cohen's κ = 1.000). We did not find any non-reproducible results (table 4).

Table 4. Reproducibility of atopy patch tests with two standardized inhalant allergens

	APT with standardized inhalant allergens			
	Pos./Pos.	Neg./Neg.	Pos./Neg. Neg./Pos.	**Cohen's κ**
Dermatophagoides pteronyssinus	39/131 (29.8%)	92 (70.2%)	0	**1.000**
Mixed grasses	5/131 (3.8%)	126 (96.2%)	0	**1.000**

G. Vehicle of Food Allergens Used in Atopy Patch Testing

Attempts at standardization of testing substances with food allergens have been made, including the development of an optimal vehicle [34], because the vehicle is obviously critical for the APT. The most frequently used are normal isotonic saline solution (0.9% NaCl) or petrolatum, but also the use of 90% dimethyl sulphoxid, cream, petroleum, petrolatum oil and methylcellulose hydrogel was described. Darsow et al. [34] compared the outcome of aeroallergen-APT with the allergen preparation in petrolatum or hydrogel. 67% of positive APT reactions were elicited with allergens in petrolatum versus 33% when hydrogel was used as vehicle. Petrolatum as vehicle may lead to improved APT results on unchanged skin. The APT using the petrolatum vehicles induced in the study of Oldhoff et al. [153] a higher number of positive APT reactions and was significantly stronger relative to the APT using allergen in aqueous vehicle. According to the Position paper of EAACI [220], the use of allergen preparation in petrolatum with defined concentration of allergens expressed in protein nitrogen units (PNU), allergen units (AU) or reactivity index (RI) is recommended.

H. Negative Control

The negative control is not used in many studies [17, 60, 79, 141, 172]. However, the negative control should be applied because of risk of false-positive reactions and to exclude possible irritant reactions. As negative control are used microcrystalline cellulose [32, 92, 93, 103, 124, 125, 211, 230], applied petrolatum [42, 104, 231, 230], petroleum [82], physiologic sterile isotonic saline solution [45, 46, 79, 82, 95, 174-176, 180], glycerol saline [27, 166, 169, 240], buffer solution [153], or empty cup [8, 25, 106, 230]. Usually, no reaction was observed to the negative control. Conversely, the use of positive control would not be necessary, because the APT only rarely elicits false-negative results [143]. However, in the literature we can find also some studies with applied both negative and positive control. As a positive control were used irritants: isopropyl myristate, propylene glycol and sodium lauryl sulphate and 0.5% solution of sodium dodecylsulphate [34, 36].

Our Data

In both our study we applied besides four fresh food allergens also isotonic saline solution (0.9% natrium chloride). We did not find any positive reaction to the negative control.

I. Pretreatment of the Skin in Testing Site

Atopy patch test has an advantage that it is non-invasive test used on the skin. Indeed, the skin is at one and the same time a barrier of the organism with a local immune system and a possible route of administration via transdermal passage of drugs [9]. So allergens put on the skin can come into contact with immune cells, but at the same time, their passage through the skin can be limited by the very low permeability of the skin [202]. However, it was also demonstrated that allergen penetration is possible without irritating physical or chemical skin alteration such as tape stripping of the stratum corneum or addition of detergents to the test preparations [34].

There are many variations considering the pretreatment of the testing site before the application of the APT set. The APT had been performed on the normal, non-eczematous skin [110, 124, 125, 140-142, 172], on the normal skin after pretreatment of the skin by scarification or stripping with adhesive tape [20, 70, 133, 2004, 217, 245] and on the mild-lesional skin [133]. Mitchell et al. [133] performed APT on skin areas which were gently abraded by removing the upper layer of the epidermis without causing capillary bleeding. In this way, allergen can more easily penetrate the skin, which is also apparent after the scratching of the skin. Gondo et al. [70] succeeded in reproducing an eczematous lesion on the apparently normal skin of a patient with AD by scratching and continuous application of allergen. Another way of facilitating allergen penetration is tape-stripping with adhesive tape, resulting in a reduction of the corneal layer. Van Voorst Vader et al. [228] found a higher number of positive APTs after rigorous tape-stripping (15x) compared to 8x stripping or no stripping. If the patients with 15x tape-strippings had been omitted, 2/3 patients tested in this manner and showing specific APT reactions would have been negative and lost. However, the number of positive reactions also increased, especially after 48-hours' reading. Tape stripping leads to enhanced penetration of allergens by removing the upper layers of the stratum corneum. One disadvantage of this method is augmentation of the number of false-positive nonspecific irritative reactions. Seidenari et al. [193] reported the highest number of positive reactions after simple application of the allergen compared to pretreatment of the skin with stripping (4x), 0.02 mL dimethyl sulfoxide, 0.05 mL of 10% sodium lauryl sulphate, and slight abrasion with the scalpel. The different pretreatment techniques partially or greatly reduced the skin reactivity. An increased number of positive APTs was found after tape-stripping x10 (27/56 patients) compared to no stripping (20/56 patients) by Langeveld-Wildschut et al. [111]. No difference in the incidence and intensity of the APT reaction was found between 10 and 20 tape-strippings. Neither 10 nor 20-tape-strippings induced nonspecific reactions. Occasionally, allergens were also applied on the lesional skin. Repeated daily application of allergen on mildly eczematous skin resulted in a marked or moderate local deterioration after 5 days [149]. This was also true, although to a lesser extent, in areas which initially were clinically uninvolved. According to the Buckley et al. [24], the use of tape-stripping is useful, if APTs are to be applied, on the grounds that positive results are more likely to be observed clinically. Various kinds of chemical and physical pre-treatment of the testing sites were used in the literature to facilitate the allergen penetration into the skin: tape stripping, skin abrasion with scalpel, or addition of detergents and chemicals (sodium lauryl sulphate) [20, 70, 81, 133, 149, 153, 213, 214], but these methods have been recently abandoned because caused many false-positive and irritative reactions.

Veremis-Ley et al. [233] studied the usefulness of laser-assisted alteration of the stratum corneum to enhance allergen delivery and patient satisfaction with this procedure aimed to facilitate the allergen penetration into the skin through the stratum corneum. Chemicals allergens were applied to the laser-pretreated sites for 60 minutes. No irritant reactions were noted and patients reported no pain. All the tested subjects with previously confirmed sensitivity to various chemical substances showed positive reactions to these components on laser-pretreated sites. With further modification, laser pretreatment may improve patient convenience and decrease irritant test reactions owing to occlusion.

J. Occlusion Time and Time of Reading

There are large differences in the duration of allergen application and reading time in atopy patch testing. Most studies used a single, prolonged allergen application and reading times of 24, 48, or 72 hours. A number of studies reported also immediate reactions at 10-20 minutes (local contact urticaria) [49, 140, 141]. Acute reactions observed within 20 minutes after APT application were correlated with elevated levels of food-specific IgE [16]. Immediate reactions after 20 minutes were defined as inconclusive and further testing was stopped [140, 141, 158].

Langeveld-Wildschut et al. [111] evaluated the APT at four different time points: 20 min, 24, 48 and 72 hours. Most patients had positive responses at 24 h, persisting until 48-72 h, 7/34 patients (20.6%) started reacting at 48-h. This subgroup had a significantly lower specific IgE level than patients who started reacting after 20 min or 24-h. van Voorst Vader et al. [228] found more specific allergic reactions after a 48-h reading than a 24-h reading alone. Optimal occlusion time was studied by Rancé [166]. In 48 children with AD were performed 64 open oral challenges. Atopy patch tests with various fresh food allergens were performed under the condition of two different occlusion times: 24 hours and 48 hours occlusion period usually used in contact allergy. The results were read 30 minutes after the removal of the cups, after 24, 48, and 72 hours for the final evaluation of the test. The sensitivity, positive predictive value and negative predictive value were all better for the 48-h occlusion time than for the 24-h occlusion time. Therefore, the recommended occlusion time for APT with food allergens should be 48 hours, as for the contact allergens. The time of reading after the removal cups differed in the studies: 15 minutes [211], 30 minutes [32, 166, 169], or after 60-120 minutes after the removal of the testing cups at 72 hours from the beginning of application [192]. However, the most studies respect the recommended interval 20 minutes after the removal of APT set from the back.

With aeroallergens studies have been performed comparing different reading times as well. An occlusion of 48 hours and a reading time of 72 hours should be preferred [48, 218, 220]. In another study, the reaction were read at 24 (and reapplied), 48 and 72 hours from APT beginning [42]. Evaluation of the APT after 48 and 72 hours gave more frequently clear-cut positive reactions than after 24 hours [42]. In the study of Bygum et al. [25], the occlusion time was also 48 hours and the reading was performed 20 minutes from the beginning of the APT and then on days 2, 3 and 7 according to standard criteria, almost all positive results were observed on day 2, and nothing was gained from later readings. Some reactions became more infiltrated on day 3 (typically crescendo phenomenon seen in true allergic responses) and reactivity decreased on day 7.

It is preferable to have a reading at 42 and 72 hours also for chemicals. An extra reading on days 6 or 7 is useful as it gives additional information in 8.2% of patch-tested patients with chemicals [99]. Delayed (7-day) patch test readings are especially important in atopic patients to distinguish allergy from irritancy and to evaluate for steroid allergy [137]. For a single 2nd patch test reading after day 2, day 3 is the best day, and especially better than day 4. If a 3rd reading is performed, it should be done at day 5 to get the maximum information out of patch testing [64]. In another study, delayed reading after one week revealed redness and itching at the area of cup placement only in the small amount of the tested patients positive on the reading at 72 hours. The authors concluded that this additional reading does not add any additional information to the classic and recommended final evaluation at 72 hours from the beginning of the test [106].

According to the standardization of this procedure, occlusion time should be 48 hours. The reaction sites should be checked after 20 minutes from the beginning of the testing for the presence of immediate-type local contact urticaria (IgE-mediated skin sign). The later evaluation of the test is 20 minutes after the removal of the cups and the final evaluation is done at 72 hours from the beginning [220].

K. Scoring Systems of Atopy Patch Test Reactions

The right evaluation of the eczematous reaction occurring during atopy patch testing is the most important step in the clinical use of this test. The interpretation of the APT outcome should be done by experienced physician in the context of clinical symptoms. The clinical meaning of various grades of APT skin reaction was investigated.

The scoring systems differed between the studies. Isolauri et al. [92] scored their results as negative, irritant, significant erythema, erythema with oedema and eczema. Scoring system according to Breneman et al. (1989) envolves: "Tr" = minimal erythema, "1" = significant erythema, "2" = significant erythema plus edema (palpable elevation), "3" = significant erythema plus significant edema, early vesicle, and "4" = significant erythema plus significant edema plus vesicle and bullas [16]. The results of Osterballe et al. [158] were classified as (+) = weak positive reaction: erythema and slight infiltration; (++) = strong positive reaction: erythema, infiltration and papules; (+++) = very strong reaction: erythema, infiltration, papules and vesicles. A positive outcome of the APT was defined as ≥ + with infiltration as the major criteria. Another possibility is erythema = +, erythema, edema = ++, erythema, edema, blisters, or papules = +++, confluent blisters = ++++ [201]. Today, the standardized reading of the APT reactions is according to the International Contact Dermatitis Research Group (ICDRG) or European Task Force on Atopic Dermatitis (ETFAD) [105, 220].

The interpretation of APT to foods is not fully standardized. Some studies were aimed to validate the reading of the APT in terms of diagnostic accuracy of individual skin signs. In the study of Heine et al. [80], the skin reactions were described and graded for erythema, induration, papule formation and crescendo phenomenon (increase of skin sigh severity from 48 to 72 hours). The values of sensitivity, specificity, PPV and NPV were calculated for each skin sign in comparison with the outcome of DBPCFC. The combination of any skin induration plus papules (seven or more), or of moderate erythema plus any induration plus seven or more papules had a positive predictive value and specificity for challenge outcome of 100%, however, sensitivity was low (8% and 15% respectively). The best diagnostic accuracy for single skin sign was found for induration beyond the Finn Chamber margin (PPV 88%, SP 99%, SE 9%) and presence of at least seven papules (PPV 80%, SP 96%, SE 21%). Presence of both induration and of at least seven papules at 72 hours was the APT skin signs with the greatest diagnostic accuracy for food allergy in children with AD. The close association between induration or papule formation and positive DBPCFC appears plausible and most likely reflects skin infiltration by lymphocytes and other pro-inflammatory cells in response to prolonged epicutaneous exposure to food allergens. Because of use of the non-standard testing substances, the results of APT should be interpreted with a great care to avoid both false-positive and false-negative reactions, and it is important to test a suitable number of control subjects. If the test gives positive result, it has to be demonstrated that the actual test preparation is non-irritant and non-toxic in healthy control subjects, otherwise the observed reaction does not prove allergenicity [79]. Careful and standardized documentation of the severity of single APT skin signs may improve the reproducibility of APT interpretation and reduce intra- and interobserver errors [80]. The cutaneous reactions can be photographed and saved for the future.

Atopy patch testing involves placing allergens on the skin on day 1, removing the applied allergens after 48 hours, and interpreting the cutaneous reactions. The first reading is performed after 20 minutes after the removal of the allergens and second reading is done at 72 hours from the beginning of the testing. The first and second readings are essential for deciding whether the reaction to an allergen is more likely to be irritant (the reaction usually decreases between the two readings, decrescendo phenomenon) or allergic (the visual score usually increases between the two readings, crescendo phenomenon). The relevance of the APT outcome is determined in the context of patient's history and other diagnostic tests. A positive allergic APT reaction is "generally interpreted to be" a reaction manifested by erythematous papules, vesicles, or a spreading reaction with crust and ulcerations. According to nearly all studies, macular erythema is considered to be "questionable", and finally negative result of the APT [44]. It has been reported several times, that true allergic reaction seen on the skin under the area of occluded cups tends to grow stronger and more evident between 48 and 72 hours [103, 124, 140, 230]. Predictors for positive APT reaction with aeroallergens are these: elevated serum specific-IgE, positive SPT with the same allergen, elevated total serum IgE, the duration of the AD, and allergic rhinoconjunctivitis [43].

The interpretation of the APT is still subjective and prone to intra- and interobserver variation [42]. Therefore, many attempts have been made to standardize the methodology of APT [34, 63, 81, 105, 153, 220].

The scoring system and the interpretation of the APT results have been carried out in many different ways by different investigators. Apart from the fact, that many investigators use their own visual scoring system, visual assessment of the APT reactions might be also biased by the investigator's subjective interpretation [23]. A possibility to improve the interpretation of patch test reactions would be the use of non-invasive measurements. Heinemann et al. [81] compared the utility of chromametry and laser Doppler imaging in combination with classic visual scoring. Comparing these three methods, visual scoring was superior to both objective methods in differentiation between irritative and allergic reactions. According to the results of this study, it seems to be no practical benefit in the assessment of allergic reactions by chromametry or laser Doppler perfusion imaging. The same internationally accepted visual scoring system should be used by all investigators to allow better comparability of reaction assessment in different laboratories. Several authors criticized the visual reading of skin test reactions as subjective, of poor reproducibility, lacking in sensitivity and highly variable between observers and/or institutions. In consequence, several instrumental methods of assessment have been strongly promoted and do indeed offer several advantages, not least their objectivity. Basketer et al. [4] studied double-blinded scoring of irritant reactions by different pairs of trained observers. The grading patterns produced were almost identical, statistical analysis showed that properly trained observers were in fact able to reliably measure a grade of erythema to within ±1 on a 10 point scale; 97.6% of scores were within 2 grade points on this scale. These results provided the evidence that visual scoring could be sensitive, reliable and reproducible within a testing institution. Bygum et al. [25] observed a dose-related increase in transepidermal water loss and erythema at test sites in patients with AD and positive APT compared to those with a negative APT and controlled. The measurements of capacitance could not distinguish between positive and negative test. In the study of Zuang et al. [247], the quantification of the APT outcome was evaluated using laser Doppler velocimetry, transepidermal water loss and colorimetry. The Doppler velocimetry showed the highest correlation with the severity of the reaction. The best prediction model was obtained when the data of the three instruments were considered together. Assessing the visual score is superior to these methods. All these three methods could help in the harmonization of APT reactions in different laboratories, thus allowing a more homogenous interpretation of these reactions. However, the cut-off of positive APT reaction needing to be at least infiltrated (not only erythema) has been shown previously many times [39, 195]. A visual score was recently shown to be superior in differentiation between irritative and allergic responses compared with chromametry and laser Doppler imaging [81]. The advantages of non-invasive measurements are the objectiveness and suitability for dose-response analysis, but none of these techniques can replace the eyes and fingers of the experienced clinician [6].

Another important question is the meaning of isolated erythema. Several studies investigated the clinical mean of isolated erythema (without palpable infiltration or papules) as a final result of APT. According to the Position paper of EAACI, this outcome of the test should be considered as a questionable, finally negative result [220]. Isolauri et al. [92] considered isolated erythema as positive skin response and achieved good results in the diagnosing FA in children with AD. In another study, changing the cut-off (isolated erythema as positive result), did not improve the outcome of the test. Even the number of false-positive reactions among the control subjects increased, revealing a lower positive predicting ability, and furthermore a decrease in reproducibility was noted [79]. In the study of Saarinen [180] erythema over half of the cup and erythema with induration were considered to be positive responses. Davis and Yiannias [44] studied the clinical prevalence and relevance of reactions graded as macular erythema, which are in the most studies considered as negative outcome of APT. They examined 2 823 patients with suspected allergic contact dermatitis with standard patch tests with chemicals and some inhalant allergens. All together, they evaluated 193 530 allergen applications. On day 5, with the exclusion of reactions graded as irritant, 7 274 allergen applications were associated with positive reactions; including 42.4% graded as macular erythema. Of the macular erythema reactions, 2 430 were graded as clinically relevant. The rate of reactions in our patients was 2.2% if macular erythema was excluded, 3.8% if all macular reactions were included, and 3.4% of only those macular reactions deemed relevant were included. Patch test reactions rated as macular erythema are common and may be of clinical significance; therefore they should not be disregarded. Erythema, despite of its high specificity, appeared to be less reliable as a diagnostic criterion, as it is prone to subjective assessment. The presence of moderate erythema alone was not sufficient for a positive APT. Similarly, crescendo phenomenon was highly specific but lacked sensitivity. In combination with the other skin signs, this phenomenon increased the specificity only marginally without any changes in PPV [80]. The diagnostic properties of each skin sign are improved when used in combination. Some authors considered also isolated erythema without palpable infiltration as positive reaction [201].

L. Allergic Versus Irritant Reaction

Differentiation between allergic and irritant contact dermatitis reactions is very important for the appropriate clinical assessment of the reaction, as both these skin reactions are clinically, histologically and immunohistologically very similar. The studies in mice revealed that the chemokine IP-10 is exclusively expressed in allergic contact dermatitis. Flier et al. [59] investigated the differences in the mRNA expression of IP-10, and the related CXCR3 activating chemokines, Mig and IP-9. IP-10, Mig, IP-9 mRNA were expressed in 7/9 allergic contact dermatitis reactions after 24-72 hours, but not in the irritant reactions caused by sodium lauryl sulfate. Also the expression of ICAM-1 by keratinocytes was found only in allergic contact dermatitis reactions. Up to 50% of the infiltrating cells in allergic contact dermatitis expressed CXCR3, in contrast to only 20% in irritant contact dermatitis. This study demonstrated the differences in chemokine expression between allergic contact dermatitis and irritant contact dermatitis reactions, which might reflect different regulatory mechanisms operating in these types of skin reactions and may be an important clue for differentiation between allergic and contact dermatitis reactions. The skin responses following application of aeroallergens developed with an increasing intensity in the course of time. These reactions should be considered as true positive allergic reactions [15]. In a study performed by Niggemann et al. [140] food allergens were tested in parallel in a 1:10 dilution to exclude false-positive results by irritant reactions. The authors found positive reactions in 18 of 77 (23%) patients with the 10% diluted APT, and these were seen in the patients with the strongest reactions to the undiluted APT. Furthermore, all these patients showed late eczematous reactions in DBPCFC, mostly to wheat. The authors concluded that the APT results are not biased by unspecific, irritant reactions and that undiluted native foods should be used.

Irritation was defined as redness strictly limited to the contact area of the cup, with no infiltration, and the reaction tended to become weaker by the next examination [17, 103, 124, 125, 172, 211]. Another possible description of irritant reaction is sharply defined brownish erythema, blistering, lack of clear infiltration, decrescendo phenomenon) were considered as negative [124, 130, 166, 172]. Although APT and irritant responses share several characteristics in their outcome, the induction of these two reactions is different [113]. For the right interpretation of the APT result, it is very important to discriminate strictly between allergic and irritant reactions (figure 5) [143].

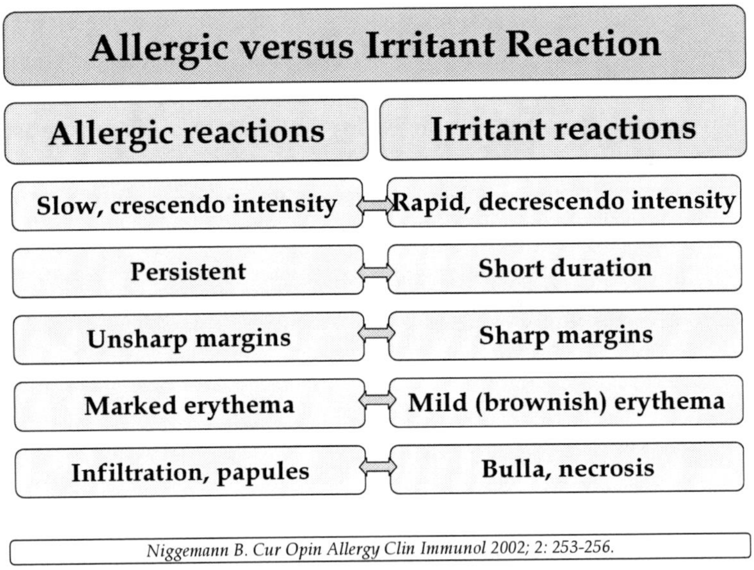

Figure 5. Allergic versus irritant reactions.

M. False Positivity and False Negativity of Atopy Patch Test Results

In food atopy patch testing, there is a high risk of false positivity, and therefore it is necessary to test in parallel also negative control and the positive results should be confirmed by elimination-challenge test [16]. Because of lower specificity and PPV, false-positive results are common, and a positive result in either SPT or APT is not sufficient to confirm cow's milk allergy. Therefore, DBPCFC is still unnecessary to confirm or exclude the diagnosis of FA in the face of positive test results [106]. In APT, there is higher risk of false-positive results, than those negative [92]. Cudowska et al. [32] found no correlation between APT and OFT in 4/34 children. The percentage of false-positive results was 25%. The percentage of false-positive and false-negative results was comparable for SPT, sIgE and APT. In another study, 5 patients with a negative challenge were positive in APT. This suggests APT is similar to traditional allergy skin testing with patients developing tolerance to the food before percutaneous testing is negative [60]. Several false-negative reactions were seen using SAFT and APT, and false-positive reaction to egg also in APT [79]. In the study performed by Stromberg et al. [211], false-positive SPT results together with false-positive APT results were uncommon. Vanto et al. [230] observed that the APT was slightly irritating the skin, causing erythema also when APT with negative control was performed. Because of positive reactions to negative control and positive reactions seemed in the subjects with negative OFC, they concluded that the APT could not be recommended as an effective diagnostic tool.

In food atopy patch testing under some conditions; also false-negative APT could be clinically relevant. Skin reactivity has been suggested as one possible cause. Evidence supports the failure to respond to a specific antigen might be due either to a faulty immune response, a defense inflammatory response or both. Thus, skin hyporeactivity may have clinical relevance in routine patch testing. Skin hyporeactivity is usually confused with skin anergy, a term used to describe an impaired cellular immune response to allergens, as seen in sarcoidosis. Since this hyporeactivity state may cause false-negative patch tests, it may have clinical relevance in routine testing. Hyporeactivity occurring during APT can be due to too low allergen dose, or not properly fixed APT set. Skin hyporeactivity could be also observed in patients on corticosteroids, immunomodulators or exposed to UV radiation. Skin hyporeactivity as one of several factors causing false-negativity should be considered to exist not only in patients with obviously immunomodulated diseases, but also in normal subjects. The synergy of several factors involved in skin reactivity would explain negative APT in truly sensitized patients and the false-positive reactions, for example, in excited skin syndrome. False-negative patch tests might reflect the impairment of an important pathway necessary for positive patch test results [107].

N. Medicament Withdrawal

It was confirmed, that various kinds of systemic or topical treatment have important and significant impact on the final outcome of the APT. According to the Position paper of EAACI, there are several exclusion criteria for APT, involving the withdrawal of topical corticosteroid therapy for at least 7 days, discontinuance of oral therapy with corticosteroids, cyclosporine A and tacrolimus and avoidance of antihistamines for at least 5 days. The patients should not have received ultraviolet treatment for at least 4 weeks [220]. The data about possible influence of oral antihistamines are missing in the literature, but as the IgE-mediated reactions play important role also in the induction of APT skin reaction, it is probable that this therapy should be discontinued according to the half life of particular preparation. No information is available concerning oral antihistamines. Considering late-phase reaction of the APT, no influence would be expected, however, erythema may be decreased. Therefore, antihistamines should be withdrawn at least 72 h prior to the APT (depending on the substance) [105, 220]. It was showed, that also the application of topical immunomodulators (tacrolimus, pimecrolimus) has significant influence on the test results, so it is necessary to avoid also this kind of topical treatment [41, 105, 220]. For general precautions, patients should neither be in an acute flare-up stage of AD, nor should pregnant patients be tested [105, 219].

The data about the withdrawal of systemic and topical medications are usually missing in the published studies. Several authors discontinued the use of all topical and systemic anti-inflammatory agents (oral and topical corticosteroids, antihistamines) for at least 48 hours [161], 72 hours [17, 80], 5 days [42] or 1 week [25, 34, 166, 201]. Bygum et al. [25] recommended the patients to avoid the sun-bathing for at least one month prior to testing.

O. Modulation of Atopy Patch Test Results with Medicaments

The APT is a reproducible model for the study of allergic inflammatory reactions in atopic dermatitis. It can be also used as a model for monitoring the capacity and effect of topical treatment. Several studies have investigated the possible modulation of the APT by an anti-inflammatory skin treatment.

Langeveld-Wildschut et al. [114] investigated the macroscopic and microscopic effects of skin pretreatment with topical glucocorticosteroids (triamcinolone acetonide 0.1%) and tar (pix liquida 10%) on the APT reaction in AD patients. Treatment with both corticosteroids and tar was able to reduce the macroscopic outcome of the APT reaction. Furthermore, both treatment modalities had an almost equally inhibiting effect on the influx of T-cells, eosinophils and $CD1^+$, RFD^+, IFN-γ^+, and IL-4^+ cells, as well as on the percentage of vessels expressing the adhesion molecules vascular cell adhesion molecule 1 and E-selectine in response to epicutaneous aeroallergen challenge. Although both pretreatments significantly reduced the various cellular constituents of allergic inflammation, all cell types remained present. In addition, this study showed that the APT can be used to evaluate the effect of topical anti-inflammatory treatments on allergic inflammation in AD patients. APT has also been employed as a model for the initiation phase of AD in order to assess whether pretreatment of non-lesional skin with fatty acid-rich emollient has a prophylactic effect in patients with AD [10]. Pretreatment of the APT sites with emollient either prevented or diminished the development of eczema, as compared with untreated control test sites in the same patients. These results indicate that the use of fatty acid-rich emollients prevents the development of atopic eczema. The APT can be used to assess the capacity of a given regimen to exert prophylactic effects in this disease [10]. The effect of 1% pimecrolimus cream on the APT and SPT was studied and it was observed that the pretreatment with 1% pimecrolimus cream has a potential to suppress the development of lesions induced by aeroallergens exposure in patients with AD. This pretreatment also decreased the intensity of the APT reaction on the re-testing [237]. Oldhoff et al. [155] investigated the potential effect of monoclonal antibody to IL-5 – Mepolizumab on the APT in the AD patients. Mepolizumab treatment was given at days 0 and 7 in a double-blind placebo-controlled trial in 43 patients. The Mepolizumab-treated group showed no significant reduction in macroscopic outcome of the APT. There were no changes neither on the microscopical level between two studied groups. This therapy cannot prevent the eczematous reaction induced by the APT. In another study of the same group [154], the skin before the APT was pretreated once daily for 3 weeks with another topical immunomodulator – 0.1% tacrolimus, cetromacrogol ointment (placebo) or triamcinolone acetonide 0.1% ointment (positive control). Pretreated with tacrolimus did not suppress non-lesional skin infiltrate in contrast to triamcinolone acetonide. Furthermore, tacrolimus did not inhibit the induction of the APT macroscopically. An equal influx of immunocompetent cells (T-cells, eosinophils, dendritic cells, $CD64^+$ and $Fc\varepsilon RI^+$-cells) were present compared with placebo. All cell types were significantly inhibited in triamcinolone acetonide-treated skin compared to placebo. The conclusion was that pretreatment with tacrolimus 0.1% ointment does not inhibit the APT reaction in patients with atopic dermatitis. Future studies are necessary to define the preventive potential of tacrolimus on non-lesional skin in AD.

Atopy patch tests can be used to evaluate the therapeutic and protective effect of the topical anti-inflammatory treatments. To get clinical relevant APT results, very important is the withdrawal of topical anti-inflammatory drugs before the APT testing to avoid the false-negative reactions. It is therefore recommended that the topical corticosteroids, tar and topical immunomodulators should be withdrawn for at least 2 weeks before the APT is performed. The practical consequence is that the APT should be performed on skin with no previous local treatment [220].

P. Standardization of Atopy Patch Test

The methodology of epicutaneous testing with aeroallergens and standardization of test material have been extensively worked out in recent years by the group of Ring and Darsow [40], but the standardization of APT with foods needs still much work before it can be recommended for routine clinical diagnosis. The weakness of APT with aeroallergens is the lack of a provocation test: the sensitivities and specificities are calculated by comparison with clinical symptoms or clinical history, which are not enough reliable. In food allergy the oral challenge test makes the cornerstone of diagnostics, but as a test it is not completely standardized and therefore the sensitivity and specificity numbers are not always comparable [218].

Atopy patch test with food allergens may increase the identification of food allergy in patients with AD in the following cases [220]:

- Suspicious of food allergy without predictive specific IgE levels or a positive SPT;
- Severe and/or persistent AD with unknown trigger factors;
- Multiple IgE sensitizations without proven clinical relevance in patients with AD.

In Europe, the efforts for the standardization of aeroallergen APT are coordinated in the European Task Force on Atopic Dermatitis (ETFAD) which has also performed and published an extensive multicenter trial [42]. A novel technique for APT with cow's milk in a commercially available test kit was recently introduced in France (Diallertest®) and compared with classical APT. The results showed good specificity and sensitivity with no side effects [101]. However, to date, the APT with foods is not well standardized and different methods in preparing the test material are likely to cause controversial results. Most studies with foods have been performed with cow's milk, hen's egg, and wheat flour. Until validation data are available, fresh food should be preferred for testing over commercial extracts, what was showed also in our study [176]. For the future, the use of recombinant proteins, some of which are available, might be interesting [218]. Atopy patch testing with standardized foodstuffs of known allergen contents would greatly improve the quality of future studies of this kind. This would also facilitate comparison between different studies and centers, and enable follow-up of reactivity in individual patients over time [211]. Further investigations of diagnostic capacity of the APT in patients of different age groups and clinical symptoms are necessary. Studies of the repeatability of tests in the same patient and trials with duplicate tests should also be performed. A standard technique for APT should be established more detailed and criteria for positive results agreed upon. The solvent that the best enhances allergen penetration should be found. Test materials with standardized known composition and major allergen content should be available and their diagnostic accuracy at different concentrations documented. Test materials, such as fresh food and freeze-dried extracts should be investigated to see which material gives the best response, and this should be standardized [211].

During the atopy patch testing, the skin should be clean, healthy, and free of ointments, lotions, powders, acne, dermatitis, scar, hair or any other condition that may interfere. During the application, the patient should stand or sit in a relaxed position with the back bent slightly forward. The prepared patches should be applied to the upper back adjacent to the vertebrae. An alternative site is the outer surface of the upper arm. Patient should refrain from exposing patch tests to excess moisture or sweat and should be return for patch test removal after 48 hours. It is useful to mark the strip or chamber location prior to removal. After the removal of the patches, the good occlusion should be checked (the ring-shaped depression around each test). The reaction should be read not less than 20 minutes after removal. A second and final skin reaction reading should be performed at 72 hours after patch test placement. Observing the changes in skin reactions may help differentiate allergic from irritant reactions. Very useful is the reading with provided Finn Chambers reading plate, which helps locate test sites and identify reactions.

In some studies, the APT were performed before the initiation of the elimination diet [46], during the elimination diet before the OFC [106], or 2 weeks after the milk challenge [230]. However, it seems that the APT could be performed independently from elimination diet and should be read by experienced physicians, since some skin reactions should be assessed very detailed and carefully taking into account also the possibility of irritant reactions.

Contemporary standardized APT technique involves (Figure 6):

- Test area: upper back, healthy, non-eczematous skin;
- Test area without pretreatment or preparation (tape-stripping, scratching, scalpel abrasion, no chemicals, detergents, medicaments);
- Large aluminium cups Finn Chambers (12-mm) placed on hypoallergenic tape;
- Allergen concentration standardized in biologic units (200 IR/g), protein nitrogen units (5 000- 7 000 PNU/g) or in μg/mL (major allergen content). For foods the use of undiluted, fresh native foods is recommended;
- Occlusion time of 48 hours;

- Readout 20 minutes after the application for the presence of immediate reactions and at 48 and 72 hours from the beginning of the testing for the final evaluation;
- Evaluation of the skin reactions according to the reading key of International Contact Dermatitis Research Group (ICDRG) or European Task Force on Atopic Dermatitis (ETFAD) (figure 7).

EAACI Protocol for Atopy Patch Test

Test area: upper back, healthy, non-eczematous skin
Test area without chemical or physical pretreatment
Large aluminium cups (12-mm) on hypoallergenic tape
Allergen concentration standardized in biologic units (200 IR/g), protein nitrogen units (5 000-7 000 PNU/g) or in µg/mL (major allergen content).
For foods the use of undiluted, fresh native foods is recommended
Occlusion time of 48 hours
Readout 20 minutes after the application and at 48 and 72 hours
Evaluation of the skin reactions according to the reading key of European Task Force on Atopic Dermatitis (ETFAD)
Turjanmaa K et al. Allergy 2006; 61: 1377-1384.
Kerschenlohr K et al. Cur Allergy Asthmy Rep 2004; 4: 285-289.

Figure 6. EAACI Standardized Protocol for Atopy Patch Test.

ETFAD Reading Key for Atopy Patch Test
revision on October 15, 2003

negative	-	Negative, no visual reaction
	?	Questionable, only isolated erythema
positive	+	Erythema and palpable infiltration
	++	Erythema and ≤ 3 papules
	+++	Erythema and ≥ 4 papules/spreading papules
	++++	Erythema and vesicles

Turjanmaa K et al. Allergy 2006; 61: 1377-1384.
Kerschenlohr K et al. Cur Allergy Asthmy Rep 2004; 4: 285-289.

Figure 7. Revised Reading Key for Atopy Patch Test.

The present standardized reading key was revised according to the 2003 ETFAD Meeting protocol. This key has more options to describe the different morphology of positive APT reactions. It would seem more important to distinguish clear-cut positive reactions from negative or questionable ones, since only reactions showing papules or at least some degree of infiltration could be correlated to clinical relevance [36-38, 80, 143]. Thus only erythematous reactions without palpable infiltration are being considered as questionable and repetition of the test is recommended in these cases. For low reaction intensity (persisting oedema without papules) there is still a need for further studies to clarify clinical relevance. In the study of Heine et al. [80] with food APT in children with AD and suspected FA, the presence of both infiltration and at least seven papules had the greatest diagnostic accuracy for predicting the outcome of DBPCFC.

It is very important to take into consideration also the **exclusion criteria for atopy patch testing** to obtain representative and relevant results (figure 8) [37, 41, 105, 220]:

- Test site free of topical steroids for 7 days;
- Test site free of topical immunomodulators (pimecrolimus, tacrolimus);
- Test site without ultraviolet treatments for 4 weeks;
- Patient free of oral steroids;
- Patient free of oral cyclosporine A or oral tacrolimus;
- Avoidance of antihistamines for at least 72 hours prior the APT (according to the substance);
- Non-pregnant;
- Patient without acute flare-up stage of atopic eczema.

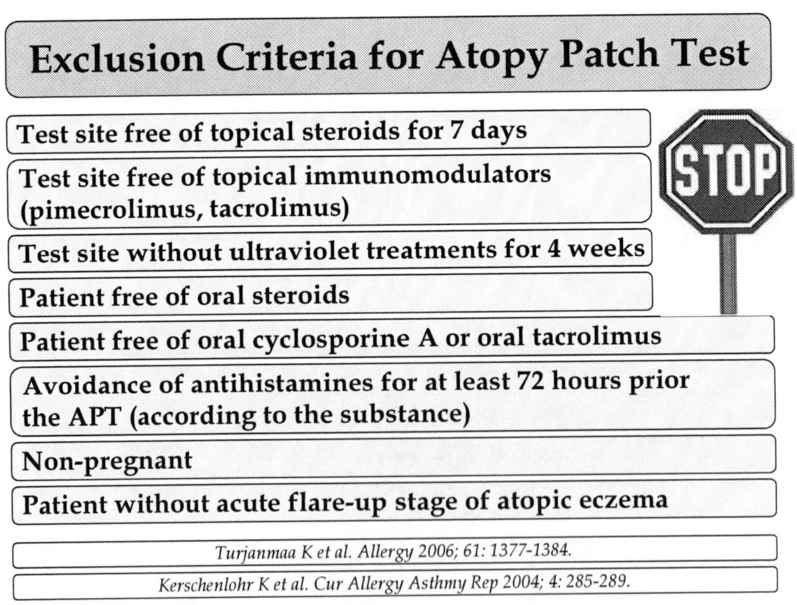

Figure 8. Exclusion criteria for atopy patch testing.

Q. Studies on Control Subjects

The major flaw of the presented studies is the lack of negative control group (non-allergic healthy children without AD), which causes the difficulty to reveal the real diagnostic sensitivity and specificity of APT used in many studies.

Osterballe et al. [158] performed APT with HE and CM in 486 Danish children with suspected food hypersensitivity. Although in the name of this paper this children population was marked as "unselected", in reality, all the children were selected according to some symptoms suspected of the relationship with FA. There was no atopic dermatitis in 381 children who could be regarded as negative healthy controls. Only three children were positive in APT with HE or CM and negative in SPT with the same allergens, and 27 subjects showed specific IgE to HE and 2 of them were also APT positive. In children with negative sIgE irritant APT reactions were present in 24/312 for CM (7.69%) and in 29/294 for HE (9.86%). The numbers of healthy, non-atopic, non-eczematic volunteers in the APT studies is usually small: 6 [140], 8 [92], 10 [34, 42, 79], 16 [98], 25 [25], 32 [192], or 85 [111]. In all of these studies, no positive reactions to aeroallergens or food allergens were observed. The majority of the studies did not include any healthy controls [8, 16, 17, 32, 60, 67, 93, 124, 125, 172, 180, 204, 207, 208, 230].

Certain numbers of positive APT with aeroallergens were reported also in patients with atopy without AD [87-91, 191, 192] and in healthy persons [27, 62, 84, 87-91, 111, 125, 192, 194] what raised the question about the specificity and the clinical relevance of positive APT results. It seems that APT with aeroallergens is more specific for AD, although it is also positive in a portion of non-AD subjects [91].

Chang et al. [28] found only 1 positive APT with CM in 15 control subjects without CM allergy. According to Isolauri et al. [92], there is a subgroup of patients with negative oral food challenge, but with positive SPT and/or APT: 14 to 19% of these patients. APT was found to be positive in 9% children in control group without CMA, although other food allergies with gastrointestinal symptoms were not excluded [46].

7. Reproducibility of Atopy Patch Tests

Before any diagnostic test is applied in routine clinical practice its results must meet the criteria for reproducibility. The studies on APT reproducibility leave important unsolved issues because of peculiar study design (including retesting at variable intervals only of subjects with positive test results), small numbers of studied subjects, variable quality and concentrations of allergens, and non-homogenous criteria for reading and reporting positive results. APT performed in parallel with the same allergen in the same individual has been found reproducible for chemical substances [12, 118, 131, 189] but the results varied from absolute agreement [66, 111, 194] to very poor reproducibility [25, 81, 87, 90]. Results from studies comparing two APT performed at different times in the same individual (overtime reproducibility) are less reproducible than tests performed at the same time in two different skin sites [22, 25, 69, 81, 91, 121, 132].

The reproducibility of the APT may dependent on several factors, such as concentration, type and number of allergens [34, 40, 111]. From classic contact patch testing with chemicals we know that the reproducibility decreased the more positive reactions have occurred simultaneously at the first test, when tested in close proximity to another positive allergic reaction ("spillover") and when a strong irritant is included in the test series [14, 134]. These can be reasons for induction of a generalized state of hyperreactivity, the so-called "angry back" syndrome or "excited skin" syndrome. In this state of hyperreactive skin, doses of substances that normally cause no reaction may lead to an inflammation. Also a subclinical inflammation where eczema is not visible can cause a hyper-reactivity of the skin which again provokes an excited skin syndrome, which may lead to positive patch tests on initial testing that are negative on retesting [134]. Reproducibility is dependent on the dose of tested allergen as well [25]. It is well known, that skin reactivity to allergens may show time-dependent variations, because new sensitizations may be acquired or lost with time. Moreover, in sensitized subjects, responses to APT may vary according to the factors modulating the local or the general immunologic state, inducing transitory skin hypo- or hyperreactivity [200], and this may influence the results and give rise to poor reproducibility when testing at different time points. Simultaneous skin contact on different areas of healthy skin should give the same response if the test material is suitable for absorption by the skin and comes into contact with the immune system. On the contrary, if the material induces an irritant reaction or is not properly absorbed by the skin, such responses may show pronounced variations. Moreover, when applying APT in children, adhesiveness to the back is not uniform because of the convexity of the scapular area and concavity of the paravertebral regions. This may also hamper allergen penetration and skin reactivity [66]. The reproducibility of APT could be influenced also by the selected squeeze dropper bottles. Selection of dropper bottles should consider drop volume reproducibility, differing drop volumes for different allergens, and the patch system being used. Important is also consistency in holding angle of the container since holding vertically produces a larger drop than holding at an angle. Important is also the volume of the tested material since it is necessary to provide appropriate volume to reach the optimal saturation of the filter paper inside the APT chambers. Clinical reproducibility of APT might increase if the dropper bottle chosen had good drop volume reproducibility and if the drop volume produced by a given bottle and test solution approximated the desired dosage for the specific patch test chamber system being used [197].

There are several studies analyzing the reproducibility of APT with aeroallergens. Heinemann et al. [81] found a reproducibility rate of only 56.3% when retesting 14 (out of 52) with a positive APT to a specific allergen. They used allergens in petrolatum with poor inter-test agreement. They tested aeroallergens on the back and retested the same allergen on the forearms 4-12 weeks later. The test agreement between the APTs with allergen from two commercial sources was poor. Bygum et al. [25] observed satisfactory reproducibility assessed by Cohen's κ statistic with the values between 0.63 and 1.00 retesting 7 patients with AD, 12 non-eczematous atopic subjects and 11 healthy volunteers with house dust mite with an interval of 15 days to 18 months. Darsow and Ring [43] retested patients with aeroallergens preparation in petrolatum. In 20 patients with AD after 6-12 months 18 (90%) showed again positive reaction in APT. In another 16 patients retested after 12-24 months, 15 (93.8%) were positive on retesting. In another study, retesting of 5 patients positive on the first testing with aqueous solution of aeroallergens, all the patients gave positive results on retesting [113]. Weissenbacher et al. [236] studied the reproducibility of inhalant APT with inter-test interval of 16 months. The positive APT reactions were highly reproducible, occurring in 15/16 (93.8%) patients. The test repeated on the back seemed to be more reproducible than that applied on the forearms. Memon and Friedman [131] elicited also high reproducibility rate of 90-95% in nickel allergic patients using classical patch testing on the back. Moreover, the forearms were clearly less responsive that the back. The possible explanation could be higher pressure on the patches at various application sites. Increased pressure on the back by lying in the bed may enhance a penetration of an allergen. Another reason could be a higher percutaneous absorption through back skin due to higher density of sebaceous gland and hair follicles [236]. Langeveld-Wildschut et al. [111] repeated aeroallergen APT in 5 patients after a period of 6 months and achieved again positivity in all the 5 patients, so the reproducibility rate was 100%. The reproducibility of aeroallergen APT in the subgroups of AD patients, atopic patients without AD and healthy volunteers was satisfactory, as the Cohen's κ values reached the level of fairly to highly satisfactory. Interesting was a fact, that the extract at 20% showed among these three groups higher variations in κ values than the extract of 30%, with which these values were unvarying [91]. Holm et al. [85] reached reproducibility rate of 79% with APT with *Dermatophagoides pteronyssinus* (10/13 patients). In two of the three patients without reproducible positive APT reaction, the eczema activity at the time of retesting was considerably lower than at the time for the first APT. The lack of reproducibility in those cases could be due to a stronger skin barrier at the time of re-testing, which allowed only a small amount of allergen to penetrate, not enough to provoke an eczema reaction. In patients without detectable levels of specific IgE in serum, the APT reaction may be a non-IgE mediated eczema reaction provoked by proteases in the house-dust-mite extract [52]. Furthermore, it is still not clear whether the reaction could be initiated by infiltration of T-cells as in classic contact dermatitis. Despite the low amount of IgE in the skin in those patients, it cannot be excluded that there may be a local IgE-mediated response in the skin, not detectable in serum, but until we have a method to identify and quantify allergen-specific IgE locally in tissue, this question remains unanswered [85]. When investigating the concordance of APTs with whole house-dust mites in 40 AD subjects at a second testing performed at a time interval of 15 days-18 months, reproducibility was highly satisfactory in AD patients and healthy volunteers (κ = 0.83 – 1.00). In a subgroup of non-eczematic atopic patients the reproducibility rate was only satisfactory (κ = 0.63) [87]. Giusti and Seidenari [66] studied right-versus-left reproducibility in the group of AD patients with *Dermatophagoides*. The reproducibility rate was 96.7% expressed with high value of Cohen's κ equal to 0.953 (highly satisfactory). As regards the agreement in the intensity of APT responses between the right and left sides, they found that reproducibility was also highly satisfactory (κ = 0.851). The agreement between APT with aeroallergens was similar to that of standard patch tests, and therefore they concluded that APTs may be considered sufficiently reproducible to be employed as a diagnostic testing procedure. In another study of the same authors [194], the right-versus-left reproducibility rate of duplicate APT with *Dermatophagoides pteronyssinus* and *Alternaria alternata* was 100% (κ = 1.00), so the agreement in the frequency of positive and negative responses between the right and left sides was 100% for both aeroallergens.

With food allergens there are only 4 short reports on the reproducibility in the literature [25, 79, 92, 101]. Discordant reactions in duplicate application were seen especially among the patients concerning SAFT, but also in 4/10 patients without egg allergy but with atopic dermatitis. The reproducibility of SAFT and APT were lower than in SPT. Taking into account an erythema as positive result, the number of discordant reactions in duplicate application increased [79]. Comparing the accuracy of Diallertest and classic atopy patch test in diagnosing cow's milk allergy (CMA), DuPont et al. performed their study on 49 children with already diagnosed CMA according to the results of oral food challenge. They found 44.8% positive results with Diallertest and 26.5% positive results with classic APT. The test results were concordant in 67.3% of cases [101, 136]. In the study of Bygum et al. [25] they retested the subjects with inhalant allergens and fresh CM after 6 weeks. The best results were achieved with *Dermatophagoides pteronyssinus*, grass and cat. In APT, comparing the results with commercially available testing substance with milk allergen and milk powder diluted to normal feed concentration, Isolauri et al. [92] obtained high concordance (Cohen's κ statistic = 0.86, 95% CI 0.77-0.95).

Our Data

We studied overtime reproducibility in our general population of 118 children attending two elementary schools in the north of Rome. This population was divided into 3 groups according to the inter-test interval: group 1 included 40 children (aged 8.41 ± 1.55 years; 54% males) whose APT was repeated after 7 days. 41 children from group 2 (aged 8.54 ± 1.50 years; 53.3% males) were retested after 14 days, while in group 3 (37 children aged 8.80 ± 0.92 years; 62.5% males) we repeated APT after 21 days.

In group 1, the values of Cohen's index kappa were very low for both sides of back (κ from -0.071 to 0.481 for the left side and κ from -0.053 to 0.481 for the right side). We did not observe any reproducible APT with cow's milk, hen's egg or wheat flour at any side. We found only 1 reproducible positive APT with tomato on both sides and 1 reproducible positive APT with wheat flour on the right side. The situation in group 2 was quite similar. The kappa values, as well the percentage of agreement between two tests on both sides of the back, were very low (κ from -0.071 to 0.643 for the left side and κ from -0.070 to 0.474 for the right side). Cohen's kappa of 0.643 (satisfactory agreement) was obtained with APT with wheat flour for the left side, but in reality there were just 2 reproducible positive APT results (and 37 reproducible negative results). In group 3, we observed 20-40% reproducible results on the left side (κ from 0.221 to 0.528) and 33-40% reproducible APT on the right side of the back (κ from 0.406 to 0.528) depending on the studied allergen (cow's milk, hen's egg, wheat flour). There were no reproducible positive APT with tomato in this group (κ = -0.057 for both sides). The percentage of agreement or kappa values did not significantly improve if we omitted consideration of the specific side of the back. We didn't observe any positive skin reaction to the tested negative control.

We observed two types of results indicating non-reproducibility: there were reactions positive in the first testing that became negative on retesting and vice versa, the reactions negative in the first testing and positive in the second testing. While in group 1 we obtained more non-reproducible results of the first "direction" (positive becoming negative), in groups 2 and 3 we observed the predominance of the second type of non-reproducible results (negative becoming positive) (table 5). In group 1 the number of non-reproducible results of the first "direction" (positive becoming negative) was significantly higher that those of second "direction" (negative becoming positive): $p < 0.001$ for the left and $p = 0.012$ for the reproducible results which became positive than those which were positive upon first testing and became negative on retesting: in group 2 $p = 0.002$ for the left and $p = 0.55$ for the right side and in group 3 $p = 0.003$ for the left and $p = 0.033$ for the right side (figure 9).

Table 5. Distribution of two types of non-reproducible APT results in the three studied groups

	Group 1 (retesting after 7 days)				**Group 2** (retesting after 14 days)				**Group 3** (retesting after 21 days)			
	CM	**HE**	**TO**	**WF**	**CM**	**HE**	**TO**	**WF**	**CM**	**HE**	**TO**	**WF**
Left side of the back												
1st pos. 2nd neg.	8 (20%)	4 (10%)	2 (4.9%)	3 (7.5%)	1 (2.4%)	0	0	0	2 (5.4%)	0	2 (5.4%)	1 (2.7%)
1st neg. 2nd pos.	1 (2.5%)	2 (5.0%)	0	1 (2.5%)	2 (4.9%)	6 (14.6%)	2 (4.9%)	2 (4.9%)	6 (16.2%)	6 (16.2%)	2 (5.4%)	2 (5.4%)
Right side of the back												
1st pos. 2nd neg.	7 (17.5%)	2 (5.0%)	2 (4.9%)	2 (4.9%)	1 (2.4%)	2 (4.9%)	1 (2,4%)	2 (4.9%)	2 (5.4%)	0	2 (5.4%)	1 (2.7%)
1st neg. 2nd pos.	1 (2.5%)	2 (5.0%)	0	0	1 (2.4%)	3 (7.3%)	1 (2.4%)	4 (9.8%)	4 (10.8%)	6 (16.2%)	2 (5.4%)	2 (5.4%)

CM cow's milk; HE hen's egg; TO tomato; WF wheat flour; neg. negative; pos. positive

According to our results, further investigation of the reproducibility of APT with food allergens is needed. In atopy patch testing with food it is necessary to establish an optimal allergen source, concentration of main allergen and procedure of testing material preparation that would lead to more reproducible results. If significant reproducibility of retesting at the same time or after certain period of time is established, the clinical validity of APT with food allergens will become more stable and evident. Some contribution to the improvement of APT reproducibility could add also the introduction of clinical non-invasive measurements for objective assessing the visual outcome of APT. The development of optimal testing material for food atopy patch testing is one of the final steps of the standardization process of this diagnostic procedure.

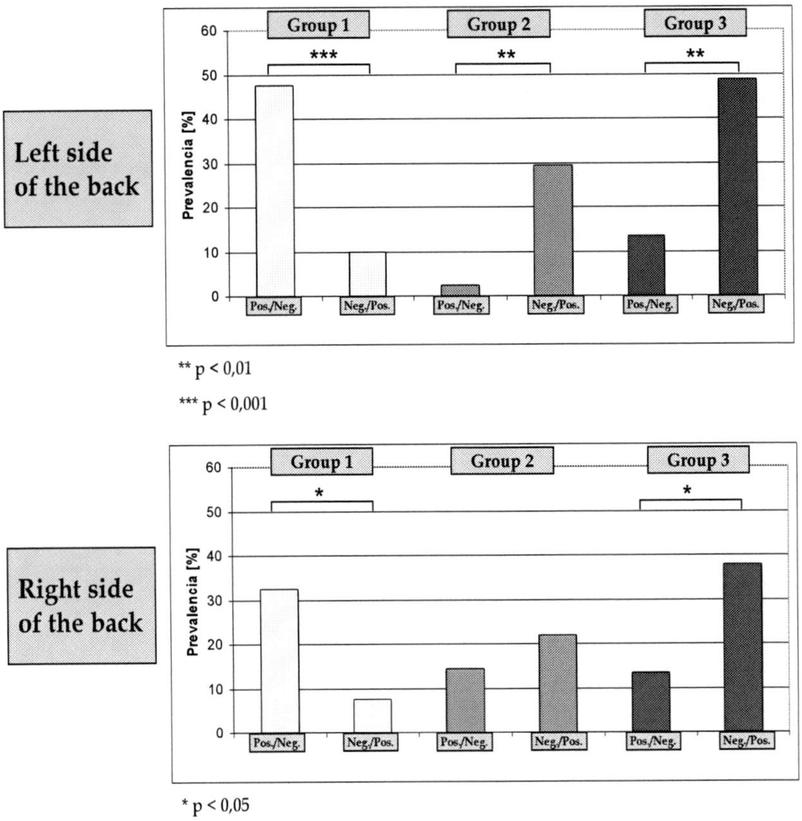

Figure 9. Differences in the prevalence of the two types of non-reproducible results.

8. Side Effects of Atopy Patch Tests

There are only a few comments on side-effects using the APT. In general, no side effects are usually observed in the clinical studies, although several sporadic reports on the side effects could be found in the literature (figure 10). In 6 children in the study of Niggemann et al. [140] developed acute local urticaria and itching (contact urticaria) 5-15 minutes after administering the APT (all with egg allergen). In 3 of 10 patients developed a systemic reaction (erythema, itching, urticaria and/or eczema outside the testing area) during or shortly after SAFT application, and this patients were consequently excluded from further testing with APT where we could expect the same kind of side effects. Other side effects of APT were severe urticaria and rhinoconjunctivitis shortly after application of APT with HE on the back [79]. In 3 subjects with proved HE allergy, the immediate type reaction was elicited by APT and the further testing was stopped [158]. No systemic reactions have been observed in the 400 patients APT tested indicating it to be a safe procedure. Only local adverse effects in patients with strong APT positivity were observed and the application of topical corticosteroids resolved these problems in 2-3 days [16]. In some children infiltration and redness of the APT area remained up to 2 months after allergen application. This was most often seen with cereals [211, 230]. During APT, few infants with strong local patch test reactions suffered from exacerbation of the present eczema elsewhere in the skin [103]. Large extensive erythematous reactions around the cup were observed by Keskin et al [106]. In another study, among 253 patients tested, 11 (4.35%) showed local eczema flare, two (0.79%) contact urticaria, two (0.79%) irritation from adhesives, two (0.79%) bronchial asthma and systemic reactions. None of the reactions was regarded as a severe side-effect [39]. In 314 patients reported by Darsow et al. [42], adverse effects were recorded in 7.7%. They were mostly mild, including local flares, contact urticaria, irritation from adhesive tapes and local itching. Conversely, some authors did not record irritation reactions to the adhesive tape used for application of aluminium cups [8]. Several authors did not observe any side effects during the APT application [101, 123, 124, 136, 166, 172].

It is possible, that with some allergens (hen's egg, tomato) the side effects could be observed more frequently than with the others (cow's milk). Although the side effects of APT testing are uncommon, APT should not be performed if the underlying skin is inflamed [206]. In some children suffering from severe forms of AD, it is not possible to provide skin testing because of extent of eczematous flares or impossibility of medication discontinuance [92].

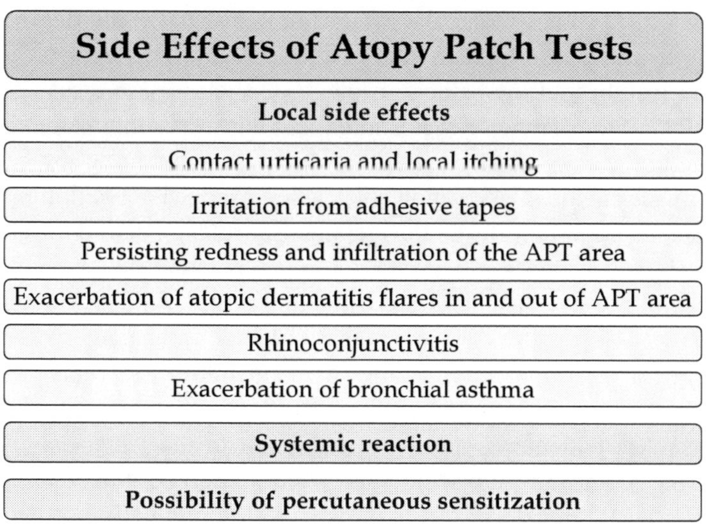

Figure 10. Side effects of atopy patch testing.

Our Data

To investigate the frequency of side effects of food APT, we analyzed APT with fresh-food allergens in an unselected children population divided into 2 groups according to the APT set used. In group A (275 children, aged 8.71 ± 1.43 years, 54.1% boys), we applied food allergens with plastic quadratic cups on hypoallergenic textile tape of 10-mm diameter (Finn Chambers, Haye's, Alphen, the Netherlands). In group B (228 children, 8.16 ± 1.74 years, 47.2% boys), we used fine blotting paper circles on transparent adhesive tape (Curatest, Lohman & Rauscher SRL, Padova, Italy) for allergen's application. We tested 4 fresh-food allergens in their native form (cow's milk, hen's egg, tomato, wheat flour). The most common side effects were contact urticaria and itching, which were observed in 2.2% of the children in group A and in 3.5% of the children in group B ($p=0.530$). In one child from group A suffering from bronchial asthma, we noticed respiratory problems that disappeared after the removal of the APT set from the back. Although we observed irritant reaction due to adhesive tape of the set only in 1.1% of the children in group A, we noted progressive irritation in 6.6% of the children studied ($p=0.002$) in group B. According to this, Finn Chambers are more suitable that Curatest for testing the food allergens [95]. In conclusion, APT is a diagnostic procedure with minimum side effects.

9. Sensitization during Atopy Patch Testing

The possibility of transcutaneous systemic sensitization during atopy patch testing is widely discussed, but only a few data are available on this topic. The skin is directly in contact with environmental molecules which are present in the air or directly in contact with the epidermis. Despite the assumption that it has a barrier function which could prevent the penetration of molecules, the skin is permeable to all substances from the low molecular weight xenobiotics to the high molecular weight proteins. Only the degree of permeability varies depending on the physiological state of the skin and chemical properties of applied molecules. Recent insight into the pathophysiology of allergic skin diseases have shown that allergen penetration is not the major factor in explaining why some patients become allergic while others maintain an immunological tolerance to the penetrating molecules. Indeed, the functional properties of some allergic molecules able to induce activation of innate immunity appear to be far more important in the development of allergy than their ability to penetrate the skin easily. However, the main group of atopy patch tested patients represents AD subjects. The principle cutaneous alterations concern the epidermis and in particular the horny layer and are responsible for the dry skin (xerosis) which is characteristic of the condition. The stratum corneum in AD patients is finer than in normal subjects and contains fewer intercellular lipids. The lamellar organization of the corneocytes is altered. The alteration of the skin barrier in AD patients is shown by reduction in the water-content of the stratum corneum and by the increase in transepidermal water loss. These anomalies are observed in the inflammatory zones, but also in the skin areas with a clinically normal appearance. The alterations of this barrier are considered by some as a primitive, genetically determined abnormality (innate alteration) accelerated by permanent aeroallergens penetration causing chronic inflammation. This increases the quantity of allergens which are able to penetrate, resulting in a vicious circle and perpetuating the eczema lesions [5].

It is evident that during APT, it comes to the disturbance of the skin barrier [65] which is expressed and confirmed by the measurement of transepidermal water losses. Interesting findings brought the study of Bygum et al. [25] where a significant positive correlation was found between a positive APT, allergen dose and increase in transepidermal water loss and erythema. Allergens that repeatedly come into close contact with the skin over extended periods of time are especially likely to induce sensitization through the skin. Cutaneous exposure to food antigens is strongly dependent on eating habits. In children, especially infants, however, there is strong exposure of the skin to foodstuffs, even when they are breast fed. Such transcutaneous sensitization may even occur by the use of cosmetics and bath additives containing vegetable oil, such as peanut or olive oil, although such oils do not contain high amounts of the respective food sources [31]. All the applied allergens can during the testing enter the skin in sufficient amounts and in immunogenic form [182]. Protein antigens from various foods also penetrate into and through the skin in sufficient amount, a fact that is exploited for diagnostic purposes by rubbing foodstuff on the skin and that is obvious from protein contact urticaria, which is frequently caused by protein allergens in food [13]. The skin is ideally suited for the *de novo* induction of primary immune responses, as it is the outer surface of the body and is exposed to multifarious pathogens. All the food proteins penetrating the skin are able to provide all the necessary co-stimulatory signals to activate naive T-cells fully and thereby initiate primary antigen/hapten-specific immune responses. However, it has not been determined whether the skin is also able to induce specific immune responses to protein allergens relevant to atopic diseases and whether immune responses initiated in this way would also lead to allergen-specific IgE production, indicating an immune response that is typical of atopic diseases. In the animals model after epicutaneous application of allergen a significant allergen-specific IgE in blood was observed. Substantial amount of IgE was produced in the skin-draining lymphoid tissue, e.g. in the skin-associated immune system, so-called SALT (skin-associated lymphoid tissue). All these experiments clearly indicate that not only sensitization to protein allergens, but also immediate-type allergy resembling atopic immune responses can occur through the skin after epicutaneous exposition to various protein allergens. Primary immune responses to protein allergens can be induced through the skin (at least in animal models). These immune responses can lead to allergen-specific IgE production and increased total IgE levels. Cutaneous hypersensitivity reactions also result from this way of exposure. Primary sensitization to protein allergens, which are relevant to atopic diseases like aeroallergens or food allergens, can take place through the skin at least in animal models, but is likely to occur also in man. What contribution is made by cutaneous allergen exposure leading to primary sensitization, or by only secondary allergen exposure, to the induction, continuation, differentiation, and outcome of human allergic responses associated with atopic diseases in general and atopic eczema in particular requires further investigation. But already from the evidence available today, it is possible to say that cutaneous allergen exposure should be taken into consideration more seriously when allergen-avoidance strategies and preventive measures are planned, as well as when therapeutic regimens are developed. There are several possible effects of allergens after expicutaneous exposure on the immunologic response in the body (figure 11) [182].

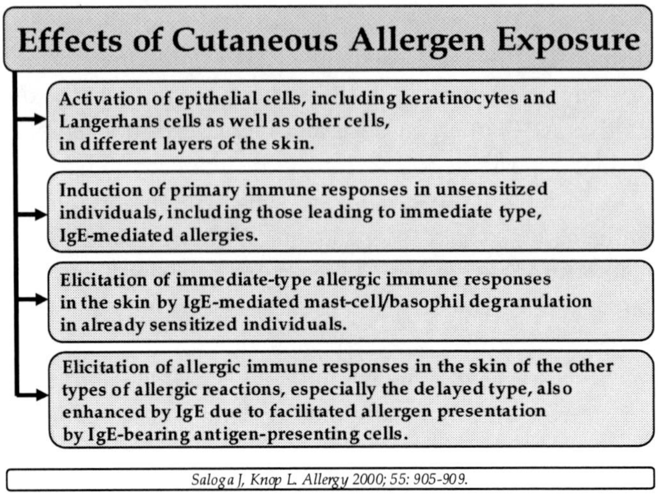

Figure 11. Effects of cutaneous allergens exposure.

It was showed, that it is possible to immunize an animal with only cutaneous exposure to high molecular weight proteins with the induction of specific IgE and the development of the inflammatory lesions of atopic dermatitis [203]. A small cohort study has suggested that the use of peanut-containing creams was more common in children who subsequently became allergic to peanut than in control infants [109]. It was also suggested, that food allergy and eczema may just represent distinct organ manifestation of a common systemic allergic response [190]. It may be that a common exposure exists for both diseases. Antigen exposures at mucosal surfaces, such as the gut, nose and lung, generally leads to immunological hyporesponsiveness [235]. In contrast, antigen exposure through barrier-disrupted skin results in strong T_H2 immunity, and may play an important role in giving rise to various allergic diseases. In the study performed by Strid et al. [209] epicutaneous exposure to peanut protein induced potent T_H2-type immunity with high of IL-4 and serum IgE. Primary skin exposure prevented the subsequent induction of oral tolerance to peanut in an antigen-specific manner. Upon oral challenge, mice became further sensitized and developed strong peanut-specific Il-4 and IgE responses. Furthermore, animals with existing tolerance to peanut were partly sensitized following epicutaneous exposure. It was demonstrated that primary sensitization can occur through the skin [83, 203, 210, 234]. Vaali et al. [229] observed on murine model of food allergy effective production of mucosal mast cell protease-1 together with specific IgE and IgG1 as a response to the epicutaneous exposure to protein allergen (ovalbumin) without immunostimulatory adjuvants.

In another study conducted by Soury et al. [202], the localization of β-lactoglobulin delivered into the skin by an innovative ready-to-use APT E-patch® was investigated. After 24 hours, 92% of β-lactoglobulin remained on the skin. The iterative skin stripping showed a 135-fold higher concentration of β-lactoglobulin in the stratum corneum than that found in the epidermis-dermis. Analysis of the solution in the receiver compartment by radioactivity assays or immunoassays indicates that intact protein did not cross the skin. Authors concluded that new ready-to-use E-patch® system allows native β-lactoglobulin to concentrate in the stratum corneum, in the vicinity of immunocompetent cells, but does not lead to its systemic delivery. Therefore, it was suggested that this delivery system creates ideal conditions for promoting a positive topical response with reduced risk of systemic anaphylactic reactions caused by the native form of the β-lactoglobulin. Small peptides can be effective immunogens when applied on to bare skin with a suitable adjuvant. These observations indicate that the skin barrier can be breached by different types of antigens that induce humoral and cellular immune responses in human [68, 159]. The diffusion of an antigen through the stratum corneum is dependent on its physicochemical properties and its molecular interactions with skin constituents. This could explain the differences in immunogenity of several antigens after their application onto bare skin [159].

Interesting was a finding done by Langeland [110]. A positive APT reaction to birch pollen was successfully transferred passively to a non-allergic-recipient, suggesting that the positive reaction may depend upon sensitizing factor (s) in the serum.

The risk of allergy to food proteins in cosmetics and topical medicinal agents is poorly understood. A clear and accurate identification of food allergens in cosmetics and topical agents are necessary. Given the hyperpermeability of infant skin, topical products containing food proteins of known allergenicity are contra-indicated in neonates, and for infants with AD, which may be associated with skin hyperpermeability [31].

There is no doubt that eczema lesions can be triggered by cutaneous exposition to allergens as patients develop eczema on contact with allergens applied as atopy patch tests. The possibility that patients can be directly sensitized by cutaneous exposition to these allergens is more controversial [5]. APT can disturb the barrier function of the skin [65] and there is a potential risk of transcutaneous allergic sensitization of the subjects tested [31, 210], it is necessary to perform this test only when it is indicated. Although the APT seems safe, some highly sensitized children cannot be tested by APT. Further studies are necessary to determine if these mechanisms play a role in the development of AD in man and if the cutaneous penetration of allergens is nor only responsible for the expression but also the induction of allergic immune responses.

Correlation of Atopy Patch Test Results to Clinical Features

It is very important to know and analyze the detailed relationships between positive APT results and various personal characteristics of tested subjects (age, gender, present disease, family history for atopy) or the results of the other tests used in the diagnostic work-up of food allergy and atopic dermatitis (measurement of food- and inhalant-specific IgE levels in serum, skin prick tests or oral food challenge, especially double-blind, placebo-controlled oral food challenge). Only through this analysis we can see the real position of the APT and clinical relevance of its results.

A. Atopy Patch Test and Specific IgE Measurement

Many studies investigated the relationships between total IgE levels or specific IgE levels in serum and the results of APT. In general, a high serum IgE level is predictive for positive APT results, which argues for a role of IgE in development of the APT lesions. However, approximately 7% of AD patients have positive APT results without specific IgE in serum or positive SPT. Especially in such patients, the APT might play the deciding role in detection of a relevant sensitization [43, 104]. The APT seems to have a better specificity than IgE methods and seems to reflect late-phase clinical reactions [140]. Positive SPT and/or high levels of specific IgE in serum are not a prerequisite for a positive APT responses as regards to the relevance of APT on the basis of the history of aeroallergen-triggered AD flares, APTs proved to have higher specificity (lower risk of false negativity) and lower sensitivity (higher risk of false negativity) than SPT and sIgE [216].

Studying APT with food allergens, Keskin et al. [106] did not observe significant differences in total serum IgE among the patients with positive or negative APT with CM. No association was found between the atopy patch test result and the presence of CM-specific IgE in serum [230]. Conversely, according to Chang et al. [28] exists a highly significant correlation between the positivity in APT and sIgE to the same food allergens (hen's egg, cow's milk, soybean milk). In the study of Darsow et al. [39, 42], the association between positive APT and sIgE to certain allergens was demonstrated, which suggests, that allergen-specific IgE have a role in the development of eczematous skin lesions after allergen contact. The children with positive APT with CM had significantly higher serum levels of CM-specific IgE than those with negative APT. This may be because of the presence of Langerhans cells bearing specific IgE in the epidermis [85, 113]. In the study of Weissenbacher et al. [236] a concordance of positive APT with the allergen-specific IgE antibody titer in 12/15 patients (80%) was observed.

There are more data in the literature describing this relationship for aeroallergens. The patients who had positive APT reactions had a statistically higher total serum IgE level and also allergen specific IgE in comparison with those subjects with negative APT results [111]. van Voorst Vader et al. [228] also found more positive APT reactions in patients with high serum IgE (> 1000 kU/L). Langeveld-Wildschut et al. [111] found that the group of APT-positive patients with aeroallergens showed a statistically significantly higher allergen-specific serum IgE levels than did the group of subjects with negative APT reactions. This association suggests an important role for sIgE in the reaction mechanism behind the APT. The elevated IgE levels in serum were associated with a higher probability of a positive APT to aeroallergens, which indicated a potentiating ability of the IgE in the APT reaction [84]. This positive correlation between positive APT results and specific IgE levels for aeroallergens was confirmed by many authors [34, 70, 81, 213, 228]. More complex relationship between APT reactivity, specific IgE and morphologic findings were described by Imayama et al. [86], who divided patients into four groups: two groups with positive APT reactions and high or low mite-specific IgE, and two groups with negative APT reactions, with or without specific IgE. It was noted that clinical morphologic findings were peculiar to three of the groups. The authors concluded that dust mite antigens may be involved in the development of skin lesions. Conversely, Gutgesell et al. [75] were not able to find any correlation between a positive APT with *Dermatophagoides* allergens and the positivity of sIgE in serum. They concluded that APT alone should not be an indicator to undertaken allergen exclusion measures in AD patients.

It seems that high level of allergen-specific IgE in serum is not mandatory for a positive APT reaction. This allows the conclusion that the APT may provide further diagnostic information in addition to patient's history and SPT and *in vitro* test results [34].

B. Atopy Patch Test and Skin Prick Test

Another skin test widely used in the diagnosis of food allergy and related disorders is skin prick test with commercially available extracts or so-called prick-by-prick test with native food allergens. The correlation between the results of both tests (skin prick and atopy patch test) has been widely studied.

In the study of Stromberg, the APT was found to be a more sensitive method than SPT in diagnosing FA in children with AD (mean age of the children was 16 months). Sensitivity (SE) and specificity (SP) of APT was 71% and 97% and of SPT 60% and 97% for egg [211]. Also Niggemann et al. [140-142] in their several studies showed that APT is a useful diagnostic tool with SE of 55%. Many children with a negative SPT have a positive APT with CM, which revealed a more sensitive method than SPT or sIgE in serum [45, 47]. In the study performed by Roehr et al. [172], APT was the best single predictive test for diagnosing CMA in children with AD. 40% of the studied subjects had positive results on all 3 tests (SPT, APT, sIgE). APT was found to be more sensitive and specific method than sIgE/SPT in diagnosing delayed food allergy in children with AD [32]. Saarinen et al. [180] found a correlation between the positivity of SPT and APT: 58% children with positive APT had also positive SPT with the same allergen. However, also in this study was a subgroup of patients with positive APT and negative SPT. In another study, the APT was found to be more positive among challenge-positive subjects (77%) than SPT (46%). Among these subjects, there was an important group with positive APT and negative SPT to the same allergen. The agreement between SPT and APT with the same allergen (egg white, and egg yolk) was from 73% to 77%. APT sensitivity proved significantly higher than that of SPT (79.6% vs. 46.2%); whereas specificity was lower (81.4% vs. 93.2%). According to the data from this study, combined SPT and APT improves screening for egg allergy in affected children, identifying 92% of those who were challenge positive. The combined use of SPT and APT further improves the detection rate, probably because several immunologic mechanisms are involved in food allergy in AD [67]. Vanto et al. [230] observed a significant association between positive reactions in the skin prick test and atopy patch test with food allergens ($p = 0.02$). Concordance between SPT and APT with hen's egg allergens was found also in another study [110]. All the strongly positive APT occurred in AD patients with a strongly positive prick test to the same allergen. In the study of Mehl et al. [130], performing 1700 single APT with four fresh food allergens (cow's milk, hen's egg, wheat, soy milk), as a single parameter, APT showed the best specificity compared with sIgE measurement, SPT, or both. Combining APT with either the SPT or sIgE measurement resulted in improved sensitivity and specificity. Decision point for sIgE measurement and for the SPT showed lower values when combined with a positive APT result. Combining all 3 parameters could not markedly improve the predictive capacity, 100% values were not found in any constellation. Children with a negative sIgE measurement and a negative SPT result (non-sensitized children) showed higher specificity values for the APT than those children with 1 or 2 positive IgE results. This interesting finding confirms the indication for the APT in IgE-negative patients. The specificity was still not high enough to meet the requirements of a replacement diagnostic test [130]. The sensitivity of SPT and APT, when used alone or in combination, was better for cereals than for CM. SPT was positive in 23% and APT in 67% of the children with positive challenge with cereals. Either SPT or APT was positive in 73% of the children with cereal allergy. SPT was less sensitive than APT in detecting cereal allergy, although it was more specific. No false-positive SPT reactions were found in cereal allergic children, whereas there were 5 false-positive APT reactions. SPT gave the best PPV, and together with APT also the best NPV. Either skin or patch test was positive in 52% of the children with CMA. APT seems to be superior to SPT in detecting cereal allergy in the children population [93]. APT showed considerably higher sensitivity over SPT, what is in agreement with the dominant delayed-type of allergic reactions usually reported [29]. In another study, sensitivity of SPT was higher than sensitivity of SAFT or APT which never exceeded 60%, whereas specificity ranged between 84-95% [79]. Considering only erythema as positive response, sensitivity of APT was improved, but specificity was reduced because of an increased number of false-positive reactions. In this study, no advantage of APT or SAFT in diagnosing egg allergy was found due to lack of reproducibility and reliability. Compared with SPT, SAFT and APT showed also systemic reactions and were less reproducible. None of these tests has any superiority to SPT [79]. Conversely, according to the results of Isolauri et al. [92], in AD we can distinguish IgE-mediated and T-cells-mediated reaction causing the onset of eczematous skin reaction in SPT and APT testing. Parallel testing with combination of SPT and APT can significantly enhance the accuracy in the diagnosis of food allergies in children suffering from AD. Cohen's kappa statistic for concordance of SPT and APT was 0.03 (95% CI -0.13-0.19), indicating that no agreement better than chance exists between the two tests with the same allergens and that the two test give discrepant and independent results. In CMA children with AD, the probability of detecting CMA was significantly higher with parallel skin testing than with only SPT or APT separately. Many children with negative SPT results but delayed-onset clinical reactions could be identified by atopy patch testing [86]. Food-sensitive patients with a delayed reaction to food challenge and negative SPT and low serum IgE concentrations showed patch test positivity and benefited from skin patch testing with dietary antigens [92]. This statement was not confirmed by Vanto et al. [230]; although this study was designed very similar to the first one and the same APT method was used.

However, there are some studies which did not confirm positive correlation between the outcomes of these two tests. Only 3 of 69 patients (4.3%) had both the skin prick and the atopy patch test positive simultaneously for CM. In most patch test-positive patients with CM, the skin prick test for the same allergen was negative [123]. Many patients with a negative skin prick test result have a positive atopy patch test to cow's milk. The APT was more sensitive method than the prick test or sIgE to detect CMA in the study population. Foods rarely produced positive results on both SPT and APT [205]. In 3 of 10 children with negative SPT, sIgE measurement and challenge-proved CMA positive APT results were observed. For soy there were only 1of 9 children with this combination of tests results [130]. The importance of using both test, SPT and APT, documents the fact that there exist a subgroup of patients suffering from various forms of FA which are negative in skin prick testing but have positive APT results [123, 124]. APT was found to be more sensitive method that the skin prick tests in the diagnosis of FA in AD children, especially those under 2 years of age. This differences were evident especially for wheat ($p<0.00001$) and rye ($p<0.00001$), for CM the statistical significance was lower ($p<0.05$) and for HE the difference was not significant. Many children with a negative SPT result have a positive APT result, especially in the case of cereals [211].

The results of several studies support the use of APT in combination with SPT in the diagnostic approach to a child with suspected FA-related gastrointestinal symptoms [8, 45,

47, 169] or in FA-related AD [92, 123, 124, 172, 195]. The combined use of APT and SPT has a high sensitivity and negative predictive value (NPV), which indicates that false-negative results are very uncommon, and that FA is highly unlikely when both test are negative. However, the definitive diagnosis of FA should rest upon the outcome of elimination diet and following food challenge [8]. Number of procedures could be eliminated by including APT in the routine clinical evaluation of patients with suspected FA-related gastrointestinal symptoms. Whereas SPT had a high sensitivity, but low specificity for predicting the outcome of OFC, APT had a high specificity but low sensitivity. Combining these two tests improves the overall predictive power [8]. APT with food proteins has been found to be helpful in diagnosing food allergy in cases where SPT or the measurement of specific IgE in serum failed [218]. With multi-allergic children adding of APT to the SPT and specific IgE estimation tests give more information for planning a wide enough elimination diet to get the skin and gastrointestinal tract symptomless in order to perform the challenge test which remains the only reliable test for food allergy. In contrast to more standardized APT methodology with aeroallergens, the sensitivities and specificities of food allergens can easily be estimated with food challenge tests [218]. APT together with SPT can be used for detecting polysensitization to multiple foods in children suffering from CMA [93, 103]. Combined SPT and APT significantly increase the chances of early detection of food allergy in children [32, 92, 123, 124, 140, 172, 211]. The combination of SPT and APT testing can identify potential causative food that might contribute to the pathogenesis of eosinophilic esophagitis [204, 208]. In food protein-induced enterocolitis syndrome all the patients had negative sIgE or SPT, but some of them (3/11) had APT positivity with causal food (fish).

The results on the concordance between the APT and SPT with aeroallergens are also contradictory. The classical IgE-mediated tests like SPT or sIgE measurement show positive reactions in the majority of patients with AD [34, 42, 81, 165, 171]. In contrast, the APT was associated with the more specific information, which patients really experienced deteriorating of AD after aeroallergen contact. Therefore, the outcome of the APT can only partially be predicted by SPT, sIgE or history, which alone, or in combination, can only be a substitute for specific provocation or allergen avoidance measures [42]. In contrast to SPT and sIgE, the APT gives additional information on another pathophysiological aspect, eczematous skin inflammation [42]. However, APT is not proposed as a single screening test in patients with AD. Patients with positive patch tests do not necessarily showed positive immediate skin tests. Non-atopic patients rarely showed positive patch test reaction to *Dermatophagoides pteronyssinus* (0.75%) [27]. APT with aeroallergens were seen in about 27% patients with AD, the most frequently with *Dermatophagoides*, especially in the severe forms of AD without complete concordance with SPT [56]. The patients with eczematous reactions after epicutaneous application of aeroallergens showed also positivity in an immediate-type skin testing [133]. In the multicenter study of Darsow et al. [42] conducted in 12 European centers, clear-cut positive APT with all SPT and IgE testing negative was seen in 7% of the patients with atopic eczema. APT, SPT and sIgE results showed significant agreement with history for grass pollen and egg white. With regard to the history, the APT had a higher specificity (64-91% depending on the tested allergens) than SPT (50-85%) or sIgE (52-85%). Positive APT were associated with longer duration of eczema flares and showed regional differences [42]. Positive correlation between SPT and APT with aeroallergens confirmed many published studies [25, 34, 71, 81, 90, 91, 110, 214]. In several studies a percentage of APT-positive subjects had a negative SPT [26, 34, 39, 62, 149, 163, 192]. No correlation between aeroallergen APT and SPT was observed [33, 56, 61, 73, 125]. The APT was more frequently positive than SPT not only in the group of AD patients, but also in the group with only respiratory symptoms. High value of APT in the patients with AD suggested that its routine might improve also the diagnosis of respiratory allergic symptoms [61].

Our Data

We investigated in our two studies on big unselected population of schoolchildren the agreement between the results of APT and SPT with fresh food allergens (cow's milk, hen's egg, wheat flour and tomato) and two aeroallergens (*Dermatophagoides pteronyssinus*, mixed grasses).

Table 6. Concordance between the results of atopy patch tests and skin prick tests with food allergens

		Atopy patch tests	
Skin prick tests (prick-by-prick method)		*Positive*	*negative*
Cow's milk	*Positive*	**0**	**5 (1.3%)**
	Negative	**29 (7.6%)**	**346/380 (91.1%)**
Hen's egg	*Positive*	**0**	**2 (0.5%)**
	Negative	**35 (9.2%)**	**343/380 (90.3%)**
Tomato	*Positive*	**0**	**8 (2.4%)**
	Negative	**14 (4.1%)**	**318/340 (93.5%)**
Wheat flour	*Positive*	**0**	**4 (1.1%)**
	Negative	**22 (5.8%)**	**354/380 (93.2%)**
All together	*Positive*	**0**	**19 (1.3%)**
	Negative	**100 (6.8%)**	**1361/1480 (92.0%)**

Table 7. Concordance between the results of atopy patch tests and skin prick tests with inhalant allergens

Skin tests with inhalant allergens				
		Atopy patch tests		
Skin prick tests		*Positive*	*Negative*	p
Dermatophagoides pteronyssinus	*Positive*	**11 (8.4%)**	**10 (7.6%)**	**0.027**
	Negative	**28 (21.4%)**	**82/131 (62.6%)**	
Mixed grasses	*Positive*	**2 (1.5%)**	**4 (3.1%)**	**0.017**
	Negative	**3 (2.3%)**	**122/131 (93.1%)**	
All together	*Positive*	**12 (4.6%)**	**14 (5.3%)**	**< 0.001**
	Negative	**32 (12.2%)**	**204/262 (77.9%)**	

In the first study on an unselected population of Italian schoolchildren no concordance emerged on between positive APT and SPT for foods in eitherchildren aged 9 or those aged

13 (table 6): none of the 100 positive APT reactions for the different food allergens concorded with SPT carried out with the same allergen. Conversely, APT and SPT for inhalant allergens yielded statistically significant concordance: about 50% of the children who had positive SPT reactions also had positive APT reactions for the same allergen (table 7) [175].

Also in the second study conducted on the population consisted of schoolchildren from Italy and Slovakia, no concordance between these two skin tests emerged for the same four fresh food allergens. There were only 5 of 319 children (1.6%) which showed the positivity of SPT and APT for the same allergens (1 for CM, 1 for HE, and 3 for wheat flour), but 111 of 319 (34.8%) were positive only in APT and negative in SPT [174].

In our two large populations of schoolchildren, we were not able to find positive correlation between positive APT reactions and SPTs for food allergens, but we observed a statistically significant association for inhalant allergens. It appears therefore that the results of APT differ according to the type of allergen tested. Given that SPT with inhalant allergens evoke a positive reaction by mean of a mechanism linked to IgE, our finding suggest that APT reaction with inhalants are also in some was produced by this mechanism, but that positive APT with food take place using other immunological mechanisms. This is in agreement with pathological findings reported in the literature. Several studies show that APT with aeroallergens are largely dependent on an IgE-mediated mechanism [42, 81, 85, 153], whereas in positive APT with foods, skin biopsies detected immunoelements attributable to all four Gell and Combs reactions [16, 85].

C. Atopy Patch Test and Oral Food Challenge

There is an increasing need to develop test instruments that will make oral food challenges superfluous in the diagnosis of FA. To reduce the need of DBPCFCs several studies investigated various combination of allergologic *in vitro* and *in vivo* tests to reliably predict the outcome of oral food challenges.

There are only few trials studying true late eczematous responses, which need 6-48 hours to develop and may occur only after repetitive ingestion of food. Over the last years, the APT has become a popular diagnostic tool for identification of food allergy in patients with AD, but the predictive accuracy remains controversial in different studies and needs further investigation in big, controlled and randomized trials. The delayed-type allergy is poorly understood. Cell-mediated reactions may be responsible for delayed-type symptoms [239]. In different clinical manifestations and reactions types of food allergy are involved distinct immunologic mechanisms. The delayed-type allergy to foods may also be an IgE-mediated allergy, even in cases with low total IgE and no detectable specific IgE to foods. The reactions may occur via high-affinity IgE receptors expressed on Langerhans and dendritic cells, leading to allergen-specific T-cell responses capable of promoting IgE production and delayed-type hypersensitivity reactions [128].

Studying the children with wheat allergy, Majamaa et al. [124] found that 20% children with positive challenge showed elevated wheat-specific IgE, 23% had positive SPT and 86% of wheat-allergic children had positive APT with wheat flour. The specificities were 0.93 for sIgE, 1.00 for SPT and 0.35 for APT. APT with cereals significantly increases the probability of early detection of cereal allergy in infants with AD and is helpful in the planning of successful elimination diet before DBPCFC. The specificity of APT was lower than that of other tests, and therefore the confirmation of the diagnosis with elimination-challenge test is still essential in patients with positive APT results. In another study of the same group, of the infants with challenge-proven CMA, 26% showed elevated CM-specific IgE, 14% had a positive skin prick test and 44% had a positive APT for CM [123]. In the study of Roehr et al. [172], for evaluating CMA, APT was the best single predictive test (PPV 95%), and the combination of a positive APT result with evidence of food-specific IgE or an APT result together with a positive skin prick test response optimized the PPV to 100%. For HE allergy, the APT was also the best single predictive test with PPV of 94%. The combination of 2 or more tests did not exceed the APT predictive value. In both CM and HE challenges, the predictability of oral challenges depended on the level of specific IgE. For wheat allergy, the APT proved to be the most reliable test and the PPV of 94% could not be improved by a combination with other allergologic tests. In distinguishing between early- and late-phase reactions, the APT as a single test showed a convincing PPV and proved to be superior to evidence of specific IgE or positive SPT results in predicting early reactions. Late-phase reactions were best predicted by a combination of APT and any level of specific IgE (PPV of 100%). In suspected HE allergy, a positive APT result showed the best results for any reaction to HE (PPV of 94%) and the combination of APT plus either sIgE or SPT produced equally good results (PPV of 94%). Evaluating wheat allergy, a positive APT was the best single predictor of reactivity (PPV of 94%). The combinations of the tests were studied to discriminate between early- and late-phase reactions. For CM, early- and late-phase reactions were best predicted by a combination of APT and specific IgE of any level (PPV of 100%). For HE, early- and late-phase reactions were equally well predicted by a combination of APT and specific IgE of 17.50 kU/L or greater or by specific IgE levels of 17.50 kU/L or greater as a single test, resulting in a PPV of 100% [172]. Sinagra et al. [201] studied a group of patients with CMA. In the whole group they found 16.7% positive APT. In the subgroup on the elimination diet, 35.5% of the patients were APT positive for CM, whereas in the subgroup of patients without elimination diet they observed only 10.6% APT. Considering three allergic tests performed in this study (sIgE measurement, SPT and APT with CM), 50% of the patients had negative all these tests, 32.8% had only one test positive, 14.5% had two tests positive and only 2.7% had all positive tests. However, these results do not allow drawing any conclusion about the possible usefulness of APT in diagnosing food allergies. 26% children with positive CM challenge were detected only by positive APT, whereas sIgE and SPT were negative. Patch testing improved the accuracy of skin testing in the diagnosis of FA in AD [103].

APT is the method with high specificity and sensitivity when investigating the existence of delayed hypersensitivity to food [211, 218]. APT reactions are usually related to the delayed-type reactions in food challenge [92, 103, 123, 124, 140,195, 211]. APT is clinically more reliable than the SPT, especially when testing with cereals [211]. Whereas immediate-type reactions (urticaria, pruritus, exanthema, vomitus, diarrhea, wheezing, sneezing) are associated with SPT positivity, delayed reactions (eczema, diarrhea, abdominal pain) are related to positive responses to APT [216]. APT is an informative and reliable diagnostic test in evaluating the delayed type allergic reactions [173]. SPT was more often positive (62%) than APT (35%) in children with immediate reactions to CM, whereas both SPT and APT were infrequently positive (13% and 40% respectively) in delayed-reaction type to CM [93]. In 7 patients, positive SPT was associated with delayed reaction in the CM challenge. The

APT was found to be as sensitive as SPT in demonstrating immediate-type hypersensitivity. It was more sensitive in detecting delayed-type reactions to cereals than SPT. Combining this two tests increases the test sensitivity in both immediate and delayed reactions. SPT was more specific than APT in demonstrating immediate-type reactions, whereas APT showed better specificity than SPT in late-phase reactions. SPT gave the best PPV for immediate reactions and APT for delayed reactions, whereas APT gave the best NPV for both types of reactions. Interestingly, SPT and APT with cereals were more frequently positive in children having only cutaneous symptoms than those with both cutaneous and gastrointestinal symptoms, although the difference was not statistically significant [93]. SPT was positive in 67% of the cases with acute-onset reactions to milk exposition, whereas APT tended to be negative. Conversely, APT was positive in 89% of children with delayed-type reactions; although SPT was frequently negative [92]. SPT was usually negative in patients with delayed onset of clinical reaction in OFC, whereas APT tended to be negative in those children with acute reactions. According to these results, in AD we can distinguish IgE-mediated and T-cells-mediated reaction causing the onset of eczematous skin reaction in SPT and APT testing. Parallel testing with combination of SPT and APT can significantly enhance the accuracy in the diagnosis of food allergies in children suffering from AD [92]. Immediate-type clinical reactions are more often associated with urticaria and skin prick positivity, whereas delayed reactions with atopic eczema and APT positivity [16, 32, 123, 124]. APT were positive in 63% children with positive results of OFC and it was more frequently positive in those with late-type reaction [17]. 25% of all positive DBPCFC were associated with negative food-sIgE and therefore food-specific T cells may play a predominant role in the pathogenesis of these reactions. Since APT lesions resemble spontaneous lesions both clinically and histologically, APT is likely to mimic the mechanisms involved into food/aeroallergens-responsive AD. 75% patients with isolated eczematous reaction in OFC were positive in APT [17]. Interestingly, the children positive in APT for CM were frequently positive for egg and wheat in the APT [32, 103]. Immediate-type reaction to CM challenge was associated with SPT positivity, while delayed-type reactions were related to patch positivity [103]. The predictive capacity of APT for immediate or delayed reaction in the OFC was identical. APT was positive in 16 from 17 positive OFC with CMA regardless the type of clinical reactions [172]. No correlation between the reaction's type and APT positivity was found, since infants with immediate, delayed, or negative challenge reactions showed consistent distribution of APT reactivity. The patch test results with CM do not seem to be related to acute or delayed challenge reactions [230]. Positive responses to APT with peanut were recorded in 19% of the patients, whereas in 12% positive SPT was observed. APT was more frequently positive in subjects with eczematous responses after challenge with respect to those with urticarial reactions [195]. Only 43% of the infants with positive OFC and diagnosed CMA were positive in APT and 40% of the positive APT was related to the negative challenge (false-positivity). The combination of four tests (SPT, sIgE, APT, eosinophil cationic protein) that produced the best overall agreement (0.73) had a sensitivity of 0.76 and a specificity of 0.67. The infants with a positive challenge not detected by this combination more often had a reaction of a delayed type with gastrointestinal symptoms [180]. Räsänen et al. [167] found that APT and lymphocyte proliferation tests were more often positive in children exhibiting delayed-type reactions, whereas the skin prick test and the basophile histamine-release test were more often positive in children manifesting immediate-type reaction to cow's milk. There is a subgroup of patients, especially those with late-onset reactions in OFC, which have negative SPT or sIgE results, but positive APT. These patients may have gone undiagnosed if APT had not been performed [106]. APT tended to be positive in infants with an immediate reaction [180]. APT with cereals is more predictive for cutaneous manifestations (urticaria, rash) than the gastrointestinal ones [93]. APT showed the value of NPV of 100% in the study with patients with suspected food-protein induced enterocolitic syndrome, so this test should be recommended in the diagnostic algorithm among these subjects. This will allow many patients to stop restricting their diet at an earlier age [60].

APT may be useful in the diagnostic work-up of CMA and in combination with other simple tests could negate the need for DBPCFC [106]. The combined used of APT and SPT has a sensitivity of 100% and NPV of 100%, showing that false-negative results are extremely uncommon and that in the population studied, CMA can be excluded if both tests results are negative [106], which in contrast to the results of Kekki et al. [103], who reported that of 54 children with positive CM OFC, 19 (35.2%) had negative SPT and APT results. Because of lower specificity (50%) and PPV (76%), false-positive results are common, and a positive result in either SPT or APT is not sufficient to confirm CMA. Therefore, DBPCFC is still unnecessary to confirm or exclude the diagnosis of FA in the face of positive test results [106]. The addition of CM-specific IgE does not improve the performance of the diagnostic methods, making this test superfluous when SPT and APT can be performed [106].

Some authors are skeptic about the role of APT in the diagnostic work-up of food allergy. In children with CMA or HE allergy proved by OFC, no reaction was predicted by APT alone. All positive reactions in OFC were however only immediate type, they did not observe delayed-type reactions. Osterballe et al. [158] revealed the difficulties in using APT in children 3 years of age. APT could not predict hypersensitivity to HE or CM not identified by SPT, histamine release or specific IgE. According to this, APT cannot be recommended in the diagnosis of hypersensitivity to HE or CM in children aged 3 years in a daily clinical practice. A study of Berni Canani et al. [8] was aimed to examine the diagnostic accuracy of APT using fresh food vs. commercially available freeze-dried purified food extracts. Among 31 patients with confirmed CMA, 20 had a positive APT with fresh food (3 in the group with an immediate reaction and 17 in the group with a delayed reaction). Only one child had a positive APT carried out with a commercial assay (this one had a positive APT also with fresh CM). Among the 19 patients with HE allergy, 16 had a positive APT with fresh food (2 with immediate reaction in challenge and 14 with a late reaction in challenge). Only one subjects had both a positive commercial APT and a positive APT with fresh HE [8]. Correctly bypassing an oral food challenge with combined testing, including APT, only between 0.5-7% (99% predicted probability) and between 6-14% (using 95% predicted probability) of children would fulfill the criteria for avoiding an OFC. APT was associated with late-type and combined-type reactions in food challenge with CM, soy milk and wheat, whereas sensitivity of APT for challenge with HE was lower for late-type reaction than for the early-type or combined-type reactions. Because of conflicting results, the APT does not seem to add diagnostic information, not even in children with late-phase clinical reactions. It was suggested that for daily clinical practice, the APT adds only a small predictive value to the standard SPT or sIgE measurement in diagnostic work-up of suspected food-related symptoms in children population [130]. In another study, Bygum et al. [25] investigated the clinical interpretation and reproducibility of APT in 48 selected young adult patients with and without AD using standard inhalant allergens and fresh cow's milk. In agreement with Osterballe et al. [158], they were not able to find any clinical relevance of a positive APT to fresh CM. Their results do not support the routine use of atopy patch tests in the evaluation of adult patients with AD [25]. One patient had a positive CM APT without clinical relevance, as oral intake of milk did not provoke the exacerbation of AD [25]. SPT demonstrates best concordance with the results of OFC followed by SAFT and APT [79].

The combination of positive APT results together with measurement of levels of food-specific IgE makes DBPCFC for suspected food allergy in some cases superfluous [140, 141, 172]. The combination of positive APT with defined levels of food specific IgE (CM $\geq$ 0.35 kU/l and HE $\geq$ 17.5 kU/l) makes the use of DBPCFC unnecessary for suspected allergies to cow's milk and hen's egg [172]. The associations of SPT/sIgE with immediate-onset reactions and APT with delayed-onset reactions confirm and stress the role of these methods in diagnosing of different type of FA [123, 125, 140, 211].

D. Atopy Patch Test and Age

A general problem is that the APT with foods (usually with HE and CM) has mostly been studied in infants and children, since FA plays a role especially in this age group, whereas aeroallergens (especially house dust mite) have been studied more intensively in adults. This presents a bias in the methodological evaluation of age as influencing factor. Although APT has been introduced into clinical practice, its diagnostic accuracy remains controversial, especially in older children and adults. Most of data concern the ability of APT in diagnosing allergic sensitization in infants in the first 2 years of life.

The age probably affects the APT results. Young children's skin is thought to be thinner and more receptive than that of older children [166]. Rokaité et al. [173] in their study observed that the prevalence of positive APT with food allergens decreases with an increasing age of the investigated children. Positive APT results were more often found in children from 6 months to 7 years of age. In infancy APTs were positive disregarding the amount of IgE in blood. The younger child was, the greater the possibility that the atopy patch test was positive. Possible explanation of these findings could be physiological characteristics of the skin (permeability of the skin, sweating, response to irritants and sensitivity to light) of an infant or a small child. It is necessary to perform APT in infants and pre school children, when total IgE amount in the blood is normal and SPT are negative. Stromberg [211] found also lower prevalence of positive APT among the children over 2 years compared to those under 2 years of age. This finding was explained by possibility that some children developed tolerance to the tested foodstuffs, or the skin perhaps became thicker so that the allergen applied did not penetrate as easily as in smaller children. In a large multicenter study of Darsow et al. [42], higher frequency of positive APT reactions to food allergens were seen in children compared with adults (wheat 15% vs. 8%, celery 12% vs. 8%) except for egg white (both 11%). The sensitivity of APT with CM, SM and wheat increased with age, but for HE, there was no significant differences in sensitivity among all age categories. A reason for this might be more sensitive skin in younger children and therefore more false-positive APT results [130]. While SPT reactivity proved to be higher in children above 12 years of age, APT positivity was more frequent in children under 6 years. APT sensitivity proved significantly higher than SPT sensitivity, in particular in children under 12 years of age [195]. The sensitivity of APT with cereals increased in the younger children but at the same time, the specificity decreased. Conversely, the sensitivity of APT with cereals decreased with increasing age with concomitant increase in specificity of this test [93]. The prevalence of positive APT results was higher in younger children with CMA [67, 106, 211].

Perackis et al. [161] performed in the group of 498 children aged from 3 to 148 months showed that the value of sensitivity, specificity, PPV and NPV concerning CM, HE and wheat does not seem to be influenced by age in infancy and childhood. APT may therefore be performed in the diagnostic work-up of food allergy in children with APT up to 12 years without unimpaired accuracy. A study performed by Niggemann et al. [142] it was found that the age of tested children had no influence on the outcome of APT, and furthermore the APT results showed equal distribution between early or late clinical reactions. The outcome of APT does not seem to be influenced by age of the children. No age correlation between positive results of OFC and APT and higher specificity of APT in older children confirm its accuracy in diagnosing delayed cow's milk allergy in all age groups of children [32]. High PPV confirms that APT might be performed in the diagnostic work-up of food allergy in children with AD up to 3 years of age with unimpaired accuracy [32]. There was no correlation between the age of AD patients and the positive reaction rate of APT [28].

Using aeroallergens in APT settings, more positive results with house-dust mite antigens were observed in children younger than 10 years [192]. Varela et al. [231] found a positive association between the younger age groups and the positivity of APT with mite antigens and delayed skin hypersensitivity, while the opposite was true for immediate hypersensitivity. APT should be a part of the protocol for assessing children with AD, particularly in the younger age groups. In general, only few studies have been published on children populations using aeroallergens [38], whereas many studies investigated the APT with aeroallergens in adults.

APT could be performed in infants as well as in older children, since it seems that age does not have a significant impact on the APT outcome [142].

Our Data

We analyzed the differences of the prevalence of positive APT reactions among various age categories. At first we compared the children from the first ($n = 262$, 53.4% boys, age 8.96±1.38 years) and second grade ($N = 200$, 48% boys, age 12.63±0.90 years) of grammar school. Whereas those younger showed lower frequency of positive SPT with inhalant allergens (mixed grasses, mixed trees, cat dander, *Alternaria alternata*), the frequencies of positive SPT with food allergens were equal in both age groups. Younger children were less atopic than older: frequency of at least one positive SPT to any allergens was 22.9% vs. 34.4% ($p = 0.010$). The frequencies of positive APT in the younger group were: 10.0% for CM, 8.0% for HE, 6.8% for tomato and 8.4% for wheat flour. These frequencies among older children were 4.1%, 10.3%, 3.2% and 5.6% respectively. The significant difference was observed for CM ($p < 0.001$) and tomato (just strong positive results, $p = 0.024$) [175]. These differences became more evident, when we subdivided the children into age quartiles (figure 12). The statistically significant differences were observed for CM (1st vs. 4th quartile: $p = 0.024$) and wheat flour (1st vs. 4th quartile: $p = 0.048$). The frequencies of positive APT with HE were distributed equally among 4 age quartiles.

Comparing boys and girls, there were no differences between the positive of SPT with either inhalant or food allergens. 8.4% of boys showed positive reaction in APT with CM and 3.6% with HE. Frequency of positive APT with tomato was 5.5% and with wheat flour 9.8%. These frequencies in girls were: 6.4% for CM, 5.9% for HE, 5.4% for tomato and 4.5% for wheat flour. Boys had statistically significant higher prevalence of positive APT results with HE (p = 0.038) and with wheat flour (p = 0.05) than girls (figure 13).

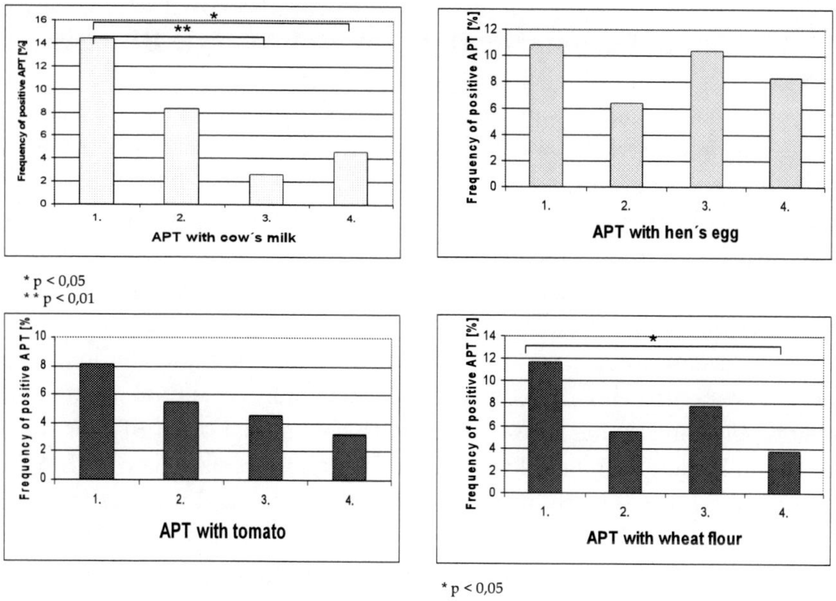

Figure 12. Frequencies of positive APT with four fresh food allergens among 4 age quartiles.

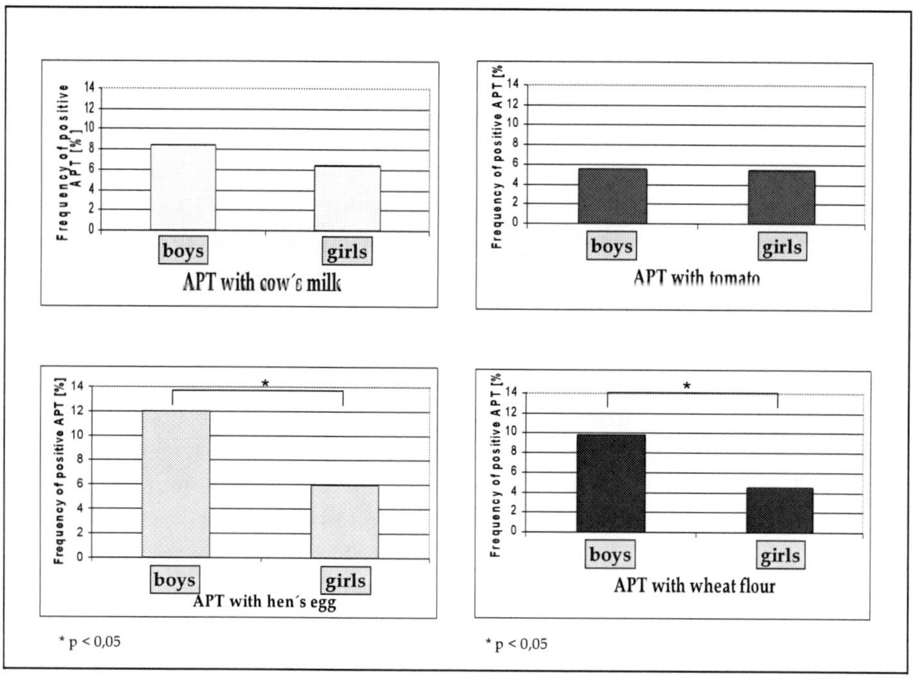

Figure 13. Frequencies of positive APT in boys and girls.

In our older study conducted in two European countries (Italy, Slovakia), we also observed more positive APT results in boys, which was the most evident for APT with HE

(19.5% vs. 8.9%, p = 0.008). Interestingly, boys showed more questionable (isolated erythema) APT results with all four tested food allergens than girls [174].

E. Atopy Patch Test in Diagnostic Algorithms of Allergic Diseases

Clinical studies have generally included patients with AD provoked by aeroallergens or associated with food allergy with or without the involvement of other systems, and the results have been conflicting [32, 92, 132, 172]. The question of whether APT identifies a subset of individuals with a distinctive clinical presentation or a distinctive immune response remains unanswered [106]. It has been repeatedly shown that in certain patients, eczematous skin lesions can be induced by epicutaneous application of aeroallergens [171, 228]. APT is frequently positive in children with atopic dermatitis [79, 93, 106, 123, 124, 141, 172, 173] and is increasingly used for detection of causal inhalant and food allergen in the pathogenesis of AD. For AD patients who do not respond adequately to initial treatment with skin moisturing, emollients and taking care with irritants, a combination of SPT or specific IgE for foods and APT can be helpful in the diagnosis of food allergies associate with AD [199, 220]. APT applying the allergens directly to the skin might be the ideal diagnostic procedure since they reproduce the characteristic inflammatory response of the disease on the affected organ itself, the skin. This method might be useful and complementary to the routine techniques of the SPT and determination of sIgE-levels [56]. Atopy patch tests, which are characterized by considerable specificity, confirm the role of polyvalent contact hypersensitivity to aeroallergens and food allergens in the development of AD. Positive aeroallergen APT results are observed in the majority of patients and can thus be regarded as an additional diagnostic criterion in AD [183]. Positive APT with food can be found in 33-86% children suffering from AD [25, 173].

The APT reactions to aeroallergens seem to be specific for sensitized AD patients, and do not occur in healthy volunteers or in patients suffering from asthma or rhinitis [18]. Conversely, in the study of Vanto et al. [230] infants with atopic eczema did not have positive patch test reactions to CM more often than those without atopic eczema (15% and 10% respectively). Positive responses in APT with aeroallergens did not occur more often in patients with atopic dermatitis, although atopy is correlated with increased skin irritability. Therefore, the APT does not appear to be suitable to distinguish subgroups of patients characterized by atopic dermatitis, asthma or allergic rhinitis [14].

It was also shown that healthy controls and patients with respiratory atopy without the history of eczema do not react to the APT [27, 34, 140] or with a lower frequency and intensity of APT reactions [195]. As no gold standard for aeroallergen provocation in AD exists, the relevance of aeroallergens for AD flares may be evaluated by APT in addition to SPT and sIgE [42]. Whitmore et al. [240] studied the prevalence of positive APT with aeroallergens in subgroups of patients with AD and/or mucosal atopy. The prevalence of reactions in patients with both atopic dermatitis and mucosal allergy (18.8%) was significantly greater than the prevalence (2.3%) in patients with only one or neither of these two atopic disorders (p=0.02).

When combining APT and sIgE results, Imayama et al. [86] classified AD patients into four groups, each group with its own distribution and morphologic features of skin lesions. Patients with an elevated specific IgE and a positive APT to house-dust-mite allergens were characterized by extensive erythematous and lichenified skin lesions and a high percentage of facial lesions (89%). Darsow et al. [35] found that APT positivity was related to the distribution pattern of eczema. The group of patients who had eczematous skin lesions predominantly on air-exposed parts of the skin such as hands, forearms, head, and neck showed a significantly higher frequency of positive APT reactions to common inhalant allergens. Two other studies [111, 113, 241] could not find a relationship between the outcome of the APT and the severity or distribution type of eczema in AD patients.

SCORAD values were significantly higher in subjects reacting to APTs and SPTs than in patients showing negative responses [196]. No significant differences in positivity figures to APT with peanut were observed between SPT positive (extrinsic AD) and SPT negative (intrinsic AD) patients (16 vs. 23%) [196]. In another study, there was no significant statistical correlation between the severity of AD assessed by SCORAD scoring system and the positive reaction rate of the APT [28]. No relationship was found between positivity of APTs and patient's history or distribution of eczema [25]. APTs with aeroallergens are more frequently positive in both "extrinsic" and "intrinsic" AD than in unaffected subjects. Other studies explaining why the APT is positive in the intrinsic form are needed, but it is probably, that this kind of AD should be marked as non-atopic allergic eczema according to the revised nomenclature [96] and the cellular immunity plays essential role in the development of this kind of eczema [88].

APT is not proposed as a single screening test in patients with AD, as it should be used in addition to SPT and sIgE as a tool to prove clinical relevance of a given sensitization. A sensitization detected by APT, which is supposedly T-cell mediated, may be even more relevant for the clinical course of atopic eczema than the demonstration of an IgE-mediated sensitization. The APT may provide an important diagnostic tool for selection those patients who show special benefit from allergen avoidance procedures or allergen-specific immunotherapy [102]. The APT with allergens in petrolatum may be used in the future as a kind of provocation test on the skin, but food challenge tests as gold standard in food allergic patients with AD are not replaced. The APT may even identify those patients with negative SPT and sIgE. However, the clinical relevance of positive APT reactions is still to be proven by standardized provocation and elimination tests and might also dependent on the APT model used and outcome definition [42]. APTs with peanuts may represent a useful integration to standard testing modalities employed for the identification of peanut allergic AD patients [196].

Seidenari et al. [192] studied three groups consisting of 72 AD patients, 40 with mucosal atopy and 32 healthy volunteers. Positive APT results were observed in 51.5% of patients with AD with mite-specific IgE, in 43.6% of patients with AD but without sIgE and in 40% of subjects affected by mucosal atopy and having sIgE to *Dermatophagoides*. From technical point of view, the simple application of the allergenic material on healthy skin gave the best results. APT with aeroallergens could contribute to a better immunoallergologic characterization both of patients with AD and patients with mucosal allergy.

No clear relationship was found between positive APT with *Dermatophagoides* and an atopic disposition of the patients or characteristics of eczema [14]. However, 64.4% of the patients with a positive APT with *Dermatophagoides* showed a response to at least 1 contact allergen of the standard series, compared to only 56.4% of the patients without a positive APT reaction ($p<0.05$). It was suggested, that some unidentified factors may contribute to positive reactions to aeroallergens that may contribute also favor an enhanced general responsiveness to contact allergens [14, 26].

It was suggested that APT works only in the patients with atopic dermatitis [18, 92, 140]. In the study of Keskin et al. [106], they observed isolated skin symptoms of FA only in 4 patients, the other with positive APT with CM had various combinations of symptoms from the part of gastrointestinal, respiratory, skin system, or had systemic anaphylactic reaction. This suggests that APT may be also useful for infants who have allergic manifestations other than AD.

The APT positivity is not specific for the patients with AD, since many studies observed also positive results in the groups of patients with e.g. gastrointestinal food allergy or respiratory allergy without skin involvement [111]. Food patch test showed good diagnostic accuracy not only for the skin manifestation of FA, but also for the signs in other organs or systems (gastrointestinal, CNS, articular, respiratory) [16]. APTs were found to be positive also among the children with FA with gastrointestinal or respiratory symptoms without skin involvement [45, 46, 60, 169, 230].

While dermatologic, respiratory, and systemic manifestations of FA are well recognized, reactions manifesting primarily in the digestive tract can be difficult to recognize, diagnose, and treat [10, 138, 151, 187]. Food allergy is now being increasingly recognized in conditions previously not labeled as "allergic", such as gastroesophageal reflux disease [138]. Non-recognition of FA may lead to inappropriate treatment and to confusion with primary gastroesophageal reflux with potentially hazardous decisions (e.g. surgery) [8]. APT may also have a role in diagnosing gastrointestinal manifestations of food allergies without skin symptoms, e.g. gastrointestinal CMA [45, 47] or patients with eosinophilic esophagitis [204-208]. It was confirmed that the APT can be used as a useful tool in the diagnostic work-up of children with food-allergy-related gastrointestinal symptoms [8].

Guler et al. [73] performed APT and SPT in 63 children suffering from asthma or allergic rhinitis. All the patients had positive SPT and high serum specific IgE levels for *Dermatophagoides pteronyssinus*. APT was performed with *Dermatophagoides* in petrolatum (200 IR/ml). Of 63 patients, 25% showed a positive APT result. APT testing may partly identify mite sensitive children with respiratory allergy. Positive APT results may imply that delayed hypersensitivity reactions play a role in children with asthma or allergic rhinitis. In patients with respiratory allergy, the positive APT results were less frequent suggesting that the involvement of delayed hypersensitivity in respiratory allergies is less important than in atopic dermatitis.

APT is useful method in detection of food allergy in children with isolated digestive symptoms without AD [45]. 24 patients with diagnosed CMA without AD were tested. APT with CM was positive in 79% children with CMA with gastrointestinal symptoms. In the patients with CMA and isolated gastrointestinal symptoms was observed good SE and SP for the APT with CM. A large use of APT in the presence of isolated digestive symptoms could improve detection of conditions related to CMA. Therefore, the sooner standardization of this testing procedure seems to be mandatory.

APT is suitable method also in patients with food protein-induced enterocolitis syndrome (FPIES) which is characterized by profuse vomiting and/or diarrhea several hours after ingestion of particular food (usually the most involved foods are cow's milk and soy milk, but also poultry, peas, fish, cereals, lentil, sweet potato, bean) [169]. APT may be used in the diagnosis of gastrointestinal allergy without evidence of IgE [169, 45, 46]. In FPIES traditional allergy testing is useless or this disorder because tests for food-specific IgE are routinely negative. FPIES is thought to be a non-IgE mediated food allergy syndrome. Although the precise mechanism of FPIES is still unclear, some studies suggested that TNF-α plays a significant role in the pathogenesis and pathophysiology of this disease [122]. The only method which can confirm this diagnosis is OFC. The APT predicted challenge outcome in 28/33 children with suspected FPIES. 16 cases were confirmed by OFC and in all of these cases APT were positive. In 5 subjects with negative OFC APT gave positive results (false-positivity). APT appears to be promising diagnostic tool in the diagnosis of FPIES [60].

APT can be used as a sensitive diagnostic tool also in eosinophilic esophagitis (EE). Eosinophilic esophagitis is a recently described disorder identified in patients with symptoms suggestive of gastroesophageal reflux disease but unresponsive to conventional reflux treatment and have normal pH probe results [179]. It was demonstrated that food allergy plays essential role in this disorder [53, 127]. Spergel et al. [204-208] in their several studies investigated the utility of combined skin testing in the identification of the causative foods in the pathogenesis of this therapy-resistant disorder. The foods most common identified by APT in these patients were wheat, corn, beef, milk, soy, rye, egg, chicken, oats, potato. Patients were positive to average 2.7±1.8 foods (range 0-7 foods). The elimination of positive foods identified by both SPT and APT led to complete resolution of the clinical symptoms in 18 of 24 patients. In addition, there was a concurrent improvement in esophageal biopsy specimens that matched the clinical improvement in these patients. Reintroduction of the same foods caused a re-emergence of symptoms in selected cases. The specificity of APT in EE was however a little bit lower than the values reported for AD [208]. Responders to the diet designed according to the results of APT and SPT were 72 of 146 patients with eosinophilic esophagitis and showed positive APT results to an average of 3.1±2.6 foods [205]. The combination of APT and SPT can identify the correct foods in 49% of the patients, not counting the patients given elemental formula because of multiple foods. If these patients are included, the percentage of specific diet success improves to 77%. In EE we need to perform both APT and SPT, because most patients had different foods that produced positive results on SPT compared with APT, suggesting a mixed immunologic mechanism [205]. It was shown that the combination of SPT and APT in the management of EE has been effective and needs to be further investigated. Elimination diets based on the positive results of SPT and APT together with milk elimination can prevent the need for an elemental diet in a majority of children with EE [208].

The skin of children with AD might be more prone to irritation and might therefore show more false-positive results. Indeed, in the study of Mehl et al. [130], specificity for CM and soy was lower in children with AD compared with that in children without AD. However, there is no explanation why this was reversed for HE. Overall, because of the lack of a uniform pattern, the author stated, that the APT does not add information for the diagnostic work-up of suspected food-related symptoms in children without AD.

Our Data

In our study, we analyzed the association between the positivity of APT with four fresh food allergens (cow's milk, hen's egg, wheat flour and tomato) and two inhalant allergens (*Dermatophagoides pteronyssinus* and mixed grasses) and questionnaire-derived incidences of atopic and non-atopic symptoms and disorders in the last year and in the personal history: headache, abdominal pain, pruritus, urticaria, nocturnal cough, cough after physical effort, nasal obturation, bronchitis, pneumonia, otitis media, allergic rhinoconjunctivitis, paryngitis, atopic eczema and asthma.

The children with positive APT to wheat flour had more frequently urticaria in the past ($p = 0.003$) or in the last year ($p = 0.036$) and cough after physical effort in the past ($p = 0.033$) or in the last year ($p = 0.019$). Children with positivity to wheat flour more frequently suffered from allergic rhinoconjunctivitis ($p = 0.031$) or eczema ($p = 0.033$) in the last year. They also had frequently bronchitis recidivans in the past ($p = 0.019$). The subjects with positive APT reactions to hen's egg suffered from allergic rhinoconjunctivitis in the past ($p = 0.020$) or in the last year ($p = 0.050$) compared to those with negative results of the APT with HE. This children also had bronchial asthma in the past ($p = 0.028$). In the subgroup of children with positive APT to cow's milk, we observed significantly higher prevalence of atopic eczema in the past ($p = 0.026$). Similarly, those with positive APT to *Dermatophagoides pteronyssinus* had more often atopic eczema in the past ($p = 0.011$). In children with positive APT with mixed grasses we observed higher prevalence of bronchial asthma in the past ($p = 0.011$) in comparison with children with negative APT results. In children with anamnestic data on the other symptoms (headache, abdominal pain, pruritus, nocturnal cough, nasal obturation) or diseases (pneumonia, otitis media, laryngitis) we were not able to detect the association with positive APT results either with food allergens or aeroallergens.

Conclusion

Atopy patch test seems to be a valuable additional tool in the diagnostic work-up of food allergy in infants with skin, respiratory and gastrointestinal symptoms, especially with regard to late-phase clinical reactions. A positive APT may help:

1) to detect clinically relevant late-phase eczematous reactions, leading to effective specific diets,
2) to prevent restrictive and unnecessary diets, which may be the consequence of misjudging late reactions by clinical assessment alone,
3) APT may be helpful in unclear and discrepant situations.

Confirmation of the diagnosis of FA is essential in patients with positive results of SPT and/or APT and for this purpose the only accurate method is double-blind, placebo-controlled food challenge followed by open food challenge. This is the definitive test for identifying food allergies and should be performed by trained allergists or immunologists on individuals with suspected food hypersensitivity.

In conclusion, it seems reasonable to suggest, that infants with a history of delayed-onset symptoms after food ingestion, negative skin prick results, and low serum total IgE concentration would benefit from atopy patch testing with a panel of dietary allergens. APT may help to prevent restrictive and unnecessary elimination diets which may be the consequence of misjudging late reactions by clinical assessment alone. Probably the most important use for APT will be diagnosing relevant food allergens in multi-allergic children (under the age of 2 years preferably) with a negative SPT and no food-specific IgE in serum.

Besides this, APT as experimental model of skin inflammation offers a unique opportunity to study the pathogenesis of AD and to test new treatment modalities. If APT becomes a part of routine pediatric practice, a lot of unnecessary procedures could be eliminated. However, further studies are needed before the APT may be used and recommended as a routine tool for diagnosis of food allergy.

Reviewed by Associate professor Milan Kuchta, M.D., Ph.D.; 2nd Department of Pediatrics, School of Medicine, University of Pavel Jozef Safarik, Kosice, Slovakia; E mail: kuchta@dnkosice.sk

References

[1] Agner, T; Serup, J. Time course of occlusive effects on skin evaluated by measurement of transepidermal water loss (TEWL). Including patch tests with sodium lauryl suplhate and water. *Contact Dermatitis*, 1993, 28, 6-9.

[2] Ale, IS; Maibach, HI. Reproducibility of patch test results: a concurrent right-versus-left study using TRUE Test™. *Contact Dermatitis*, 2004, 50, 304-312.

[3] Arslan Lied, G. Gastrointestinal food hypersensitivity: symptoms, diagnosis and provocation tests. *Turkish Journal of Gastroenterology*, 2007, 18, 5-13.

[4] Basketer, D; Reynolds, F; Rowson, M; Talbot, C; Whittle, E. Visual assessment of human skin irritation: a sensitive and reproducible tool. *Contact Dermatitis*, 1997, 37, 218-220.

[5] Berard, F; Marty, JP; Nicolas, JF. Allergen penetration through the skin. *European Journal of Dermatology*, 2003, 13, 324-330.

[6] Berardesca, E; Maibach, HI. Bioengineering and the patch test. *Contact Dermatitis*, 1988, 18, 3-9.

[7] Berardesca, E; Vignoli, GP; Fideli, D; Maibach, H. Effects of occlusive dressings on the stratum corneum water holding capacity. *American Journal of Medical Science*, 1992, 304, 25-28.

[8] Berni Canani, R; Ruotolo, S; Auricchio, L; Caldore, M; Porcaro, F; Manguso, F; Terrin, G; Troncone, R. Diagnostic accuracy of the atopy patch test in children with food allergy-related gastrointestinal symptoms. *Allergy*, 2007, 62, 738-743.

[9] Berti, JJ; Lipsky, JJ. Transcutaneous drug delivery: a practical review. *Mayo Clinics Proceeding*, 1995, 70, 581-586.

[10] Billmann-Eberwein, C; Rippke, F; Ruzicka, T; Krutmann, J. Modulation of atopy patch test reactions by topical treatment of human skin with a fatty acid-rich emollient. *Skin Pharmacology and Applied Skin Physiology*, 2002, 15, 100-104.

[11] Bischoff, S; Crowe, SE. Gastrointestinal food allergy: new insight into pathophysiology and clinical perspectives. *Gastroenterology*, 2005, 128, 1089-1113.

[12] Bousema, MT; Geursen, AM; van Joost TH. High reproducibility of patch tests [letter]. *Journal of American Academy of Dermatology*, 1991, 24, 322-328.

[13] Brancaccio, RR; Alvarez, MS. Contact allergy to food. *Dermatologic Therapy*, 2004, 17, 302-313.

[14] Brasch, J; Henseler, T; Aberer, W; Bäuerle, G; Frosch, PJ; Fuchs, T; Funfstuck, V; Kaiser, G; Lischka, GG; Pilz, B. Reproducibility of patch tests. A multicenter study of synchronous left- versus right-sided patch tests by the German Contact Dermatitis Research Group. *Journal of American Academy of Dermatology*, 1994, 31, 584-591.

[15] Brash, J; Uter, W; Dibo, M; Stockfleth, E; Swensson, O; Christophers, E. Positive patch tests with a Dermatophagoides mix relate to an increased responsiveness to standard patch test allergens. *Contact Dermatitis*, 2002, 46, 253-257.

[16] Breneman, JC; Sweeney, M; Robert, A. Patch test demonstrating immune (antibody and cell-mediated) reactions to foods. *Annals of Allergy*, 1989, 62, 461-469.

[17] Breuer, K; Heratizadeh, A; Wulf, A; Baumann, U; Constien, A; Tetau, D; Kapp, A; Werfel, T. Late eczematous reactions to food in children with atopic dermatitis. *Clinical and Experimental Allergy*, 2004, 34, 817-824.

[18] Bruijnzeel, PL; Kuijper, PH; Kapp, A; Warringa, RA; Betz, S; Bruijnzeel-Koomen, CA. The involvement of eosinophils in the patch test reaction to aeroallergens in atopic dermatitis: its relevance for the pathogenesis of atopic dermatitis. *Clinical and Experimental Allergy*, 1993, 23, 97-109.

[19] Bruijnzeel-Koomen, CA; van Wichen, DF; Toonstra, J; Berrens, L; Bruijnzeel, PL. The presence of IgE molecules on epidermal Langerhans´ cells in patients with atopic dermatitis. *Archives of Dermatological Research*, 1986, 278, 199-205.

[20] Bruijnzeel-Koomen, C; van Wichen, D; Spry, C; Venge, P; Bruijnzeeel, P. Active participation of eosinophils in patch test reactions top inhalant allergens in patients with atopic dermatitis. *British Journal of Dermatology*, 1988, 118, 229-238.

[21] Bruijnzeel-Koomen, C; Ortolani, C; Aas, K; Bindslev-Jensen, C; Bjorksten, B; Moneret-Vautrin, D; Wuthrich, B. Adverse reactions to food (Position paper). *Allergy*, 1995, 50, 623-635.

[22] Bruynzeel, DP; van Ketel, WG; von Blomberg der Flier, M; Scheper, RJ. Angry back or excited skin syndrome. A retrospective study. *Journal of American Academy of Dermatology*, 1983, 8, 392-397.

[23] Bruze, M; Isaksson, M; Edman, B; Bjorkner, B; Fregert, S; Moller, H. A study on expert reading of patch test reactions: inter-individual accordance. *Contact Dermatitis*, 1995, 32, 331-337.

[24] Buckley, C; Poulter, LW; Rustin, MHA. Immunohistological analysis of "negative" patch test sites in atopic dermatitis. *Clinical and Experimental Allergy*, 1996, 26, 1057-1063.

[25] Bygum A; Gotthard Mortz, C; Ejner Andersen, K. Atopy patch tests in young adult patients with atopic dermatitis and controls: dose-response relationship, objective reading, reproducibility and clinical interpretation. *Acta Dermato-Venereologica*, 2003, 83, 18-23.

[26] Cabon, N; Ducombs, G; Mortureux, P; Perromat, M; Taieb, A. Contact allergy to aeroallergens in children with atopic dermatitis : comparison with allergic dermatitis. *Contact Dermatitis*, 1996, 35, 27-32.

[27] Castelain, M; Birnbaum, J; Castelain, PY; Ducombs, G; Grosshans, E; Jelen, G; Lacroix, M; Meynadier, J; Mouqeolle, AT; Lachapelle, JM; Oleffe, J; Pons, A; Taieb, A; Tennstedt, D; Vervloeat, D. Patch test reactions to mite antigens: a GERDA multicenter study. *Contact Dermatitis*, 1993, 29, 246-250.

[28] Chang, DS; Seo, SJ; Hong, CK. Patch test and specific IgE levels with food antigens in atopic dermatitis patients. *Korean Journal of Dermatology*, 2002, 40, 1028-1034.

[29] Carroccio, A; Montalto, G; Gustro, N; Notarbartolo, A; Cavataio, F; D´Amico, D; et al. Evidence of very delayed clinical reactions to cow's milk in cow's milk-intolerant patients. *Allergy*, 2000, 55, 574-579.

[30] Clark, RA; Adinoff, AB. Aeroallergen contact can exacerbate atopic dermatitis: patch tests as a diagnostic tool. *Journal of American Academy of Dermatology*, 198, 21, 863-869.

[31] Codreanu, F; Morisset, M; Cordebar, V; Kanny, G; Monoret-Vautrin, DA. Risk of allergy to food proteins in topical medicinal agents and cosmetics. *European Annals of Allergy and Clinical Immunology*, 2006, 38, 126-130.

[32] Cudowska, B; Kaczmarski, M. Atopy patch test in the diagnosis of food allergy in children with atopic eczema dermatitis syndrome. *Advances in Medical Science (Roczniki Akademii Medycznej w Bialymstoku)*, 2005, 50, 261-267.

[33] Czarnecka-Operacz, M; Bator-Wegner, M; Silny, W. Atopy patch test reaction to airborne allergens in the diagnosis of atopic dermatitis. *Acta Dermatovenereologica Croata*, 2005, 13, 3-16.

[34] Darsow, U; Vieluf, D; Ring, J. Atopy patch test with different vehicles and allergen concentrations – an approach to standardization. *Journal of Allergy and Clinical Immunology*, 1995, 95, 677-684.

[35] Darsow, U; Vieluf, D; Ring, J. The atopy patch test: an increased rate of reactivity in patients who have an air-exposed pattern of atopic eczema. *British Journal of Dermatology*, 1996, 135, 182-186.

[36] Darsow, U; Behrendt, H; Ring, J. Gramineae pollen as trigger factors of atopic eczema – evaluation of diagnostic measures using the atopy patch test. *British Journal of Dermatology*, 1997, 137, 201-207.

[37] Darsow, U; Abeck, D; Ring, J. Allergie und atopisches Ekzem: Zur Bedeutung des "Atopic-Patch-Tests". *Hautarzt*, 1997, 48, 528-535.

[38] Darsow, U; Vieluf, D; Berg, B; Berger, J; Busse, A; Czech, W. Dose response study of atopy patch test in children with atopic eczema. *Pediatric Asthma, Allergy and Immunology*, 1999, 13, 115-122.

[39] Darsow, U; Vieluf, D; Ring, J. Evaluating the relevance of aeroallergen sensitization in atopic eczema with atopy patch test: a randomized, double-blind multicenter study. *Journal of American Academy of Dermatology*, 1999, 40, 187-193.

[40] Darsow, U; Ring, J. Airborne and dietary allergens in atopic eczema: a comprehensive review of diagnostic tests. *Clinical and Experimental Dermatology*, 2000, 25, 544-551.

[41] Darsow, U; Ring, J. Atopie-Patch-Test. *Hautarzt*, 2003, 54, 930-936.

[42] Darsow, U; Laifaoui, J; Kerschenlohr, K; Wollenberg, A; Giusti, F; Seidenari, S; Drzimalla, K; Simon, D; Disch, R; Borelli, S; Devillers, ACA; Oranje, AP; De Raeve, L; Hachem; JP; Dangoisse, C; Blondeel, A; Song, M; Breuer, K; Wulf, A; Werfel, T; Roul, S; Taieb, A; Bolhaar, S; Bruijnzeel-Koomen, C; Bronnimann, M; Braathen, LR; Didierlaurent, A; André, C; Ring, J. The prevalence of positive reactions in the atopy patch tests with aeroallergens and food allergens in subjects with atopic eczema: a European multicenter study. *Allergy*, 2004, 59, 1318-1325.

[43] Darsow, U; Ring, J. Atopie-Patch-Test mit Aeroallergenen und Nahrungsmittels. *Hautarzt*, 2005, 56, 1133-1140.

[44] Davis, MDP; Yiannias, JA. Should macular erythema reactions be counted as positive allergic patch-test reactions? *Dermatitis*, 2005, 17, 12-14.

[45] de Boissieu, D; Waguet JC; Dupont, C. The atopy patch test for detection of cow's milk allergy with digestive symptoms. *Journal of Pediatrics*, 2003, 142, 203-205.

[46] de Boissieu, D; Dupont, C. Patch tests in the diagnosis of food allergies in the nursing infants. *European Annals of Allergy and Clinical Immunology*, 2003, 35, 150-152.

[47] de Boissieu, D; Dupont, C. Diagnosis of non-IgE mediated digestive manifestations of food allergy. *Journal of Pediatrics*, 2004, 145, 716.

[48] de Bruin-Weller, MS; Knol, EF; Bruijnzeel-Koomen, CAFM. Atopy patch testing – a diagnostic tool? *Allergy*, 1999, 54, 784-791.

[49] de Groot, AC; Young, E. The role of contact allergy to aeroallergens in atopic dermatitis. *Contact Dermatitis*, 1989, 21, 209-214.

[50] de Vries, IJ; Langeveld-Wildschut, EG; van Reijsen, FC; Dubois, GR; van den Hoek, JA; Bihari, IC; van Wichen, D; de Weger, RA; Knol, EF; Thepen, T; Bruijnzeel-Koomen, CA. Adhesion molecule expression on skin endothelia in atopic dermatitis: effects of TNF-alpha and IL-4. *Journal of Allergy and Clinical Immunology*, 1998, 3, 461-468.

[51] de Waard-van der Spek, FB; Elst, EF; Mulder, PGH; Munte, K; Devillers, ACA; Oranje, AP. Diagnostic tests in children with atopic dermatitis and food allergy. *Allergy*, 1998, 53, 1087-1091.

[52] Deleuran, M; Ellingsen, AR; Paludan, K; Schou, C; Thestrup-Pedersen, K. Purified Der p1 and p2 patch tests in patients with atopic dermatitis: evidence for both allergenicity and proteolytic irritancy. *Acta Dermato-Venereologica (Stockholm)*, 1998, 78, 241-243.

[53] Dobbins, J; Sheahan, D; Behar, J. Eosinophilic gastroenteritis with esophageal involvement. *Gastroenterology*, 1977, 72, 1312-1316.

[54] Dreborg, S. Skin tests in the diagnosis of food allergy. *Pediatric Allergy and Immunology*, 1995, 6 Suppl 8, 38-43.

[55] Dubois, GR; Bruijnzeel-Koomen, CA; Bruijnzeel, PL. IL-4 induces chemotaxis of blood eosinophils from atopic dermatitis patients, but not from normal individuals. *Journal of Investigative Dermatology*, 1994, 102, 843-846.

[56] Echichipia, S; Gomez, B; Lasa, E; Larrea, I; Arroabarren, E; Garrido, S; Rodriguez, AJ. Epicutaneous test with inhalers in the study of atopic dermatitis. *Anales del Sistema Sanitario de Navarra*, 2003, 26 Suppl 2, 31-37.

[57] Feldman, RJ; Maibach, HI. Regional variation in percutaneous absorption of 14-cortisol in man. *Journal of Investigative Dermatology*, 1967, 38, 181-183.

[58] Fitzharris, P; Riley, G. House dust mites in atopic dermatitis. *International Journal of Dermatology*, 1999, 38, 173-175.

[59] Flier, J; Boorsma, DM; Bruynzeel, DP; van Beck, PJ; Stoof, TJ; Scheper, RJ; Willemze, R; Tensen, CP. The CXCR3 activating chemokines IP-10, Mig 3, and IP-9 are expressed in allergic but not irritant patch test reactions. *Journal of Investigative Dermatology,* 1999, 113, 574-578.

[60] Fogg, MI; Brown-Whitehorn, TA; Pawlowski, NA; Spergel, JM. Atopy patch test for the diagnosis of food protein-induced enterocolitic syndrome. *Pediatric Allergy and Immunology*, 2006, 17, 351-355.

[61] Fuiano, N; Incorvaia, C. Confronto tra skin prick test e atopy patch test nella diagnosi di allergia a Dermatophagoides in soggetti con patologia respiratoria e/o dermatite atopica. *Minerva Pediatrica*, 2004, 56, 537-540.

[62] Gaddoni, G; Baldassarri, L; Zucchini, A. A new patch test preparation of dust mites for atopic dermatitis. *Contact Dermatitis*, 1994, 31, 132-133.

[63] Gefeller, O; Pfahlberg, A; Geier, J; Brasch, J; Uter, W. The association between size of test chamber and patch test reaction: a statistical reanalysis. *Contact Dermatitis*, 1999, 40, 14-18.

[64] Geier, J; Gefeller, O; Wiechmann, K; Fuchs, T. Patch test reactions at D4, D5 and D6. *Contact Dermatitis*, 1999, 40, 119-126.

[65] Gfesser, M; Rakoski, J; Ring, J. The disturbance of epidermal barrier function in atopy patch test reactions in atopic eczema. *British Journal of Dermatology*, 1996, 135, 560-565.

[66] Giusti, F; Seidenari, S. Reproducibility of atopy patch tests with *Dermatophagoides*: a study on 85 patients with atopic dermatitis. *Contact Dermatitis*, 2004, 50, 18-21.

[67] Giusti, F; Seidenari, S. Patch testing with egg represents a useful integration to diagnosis of egg allergy in children with atopic dermatitis. *Pediatric Dermatology*, 2005, 22, 109-111.

[68] Glenn, GM; Scharton-Kersten, T; Alving, CR. Advances in vaccine delivery: transcutaneous immunization. *Expert Opinion on Investigational Drugs*, 1999, 8, 797-805.

[69] Gollhausen, R; Przybilla, B; Ring, J. Reproducibility of patch tests. *Journal of American Academy of Dermatology*, 1989, 21, 1196-1201.

[70] Gondo, A; Saeki, N; Tokuda, Y. Challenge reactions in atopic dermatitis after percutaneous entry of mite antigen. *British Journal of Dermatology*, 1986, 115, 485-493.

[71] Goon, A; Leow, YH; Chan, YH; Ng, SK; Goh, CL. Atopy patch testing with aeroallergens in patients with atopic dermatitis and controls in Singapore. *Clinical and Experimental Dermatology*, 2005, 30, 627-631.

[72] Grewe, M; Walther, S; Gyufko, K; Czech, W; Schopf, E; Krutmann, J. Analysis of the cytokine patterns expressed *in situ* in inhalant allergen patch test reactions of atopic dermatitis patients. *Journal of Investigative Dermatology*, 1995, 105, 407-410.

[73] Guler, N; Kirerleri, E; Tamay, Z; Ones, U. Atopy patch testing in children with asthma and rhinitis symptoms allergic to house dust mite. *Pediatric Allergy and Immunology*, 2006, 17, 346-350.

[74] Gunther, S; Reiser, K; Darsow, U; Wollenberg, A. Demonstration of a Clinically Relevant to house Dust Mite Allergens in Prick and RAST Negative Intrinsic Atopic Dermatitis Patients with the Atopy Patch Test. *31st Annual Meeting of the European Society for Dermatological Research,* Abstract 302, Stockholm, 2001.

[75] Gutgesell, C; Seubert, A; Junghans, V; Neumann, Ch. Inverse correlation of domestic exposure to *Dermatophagoides pteronyssinus* antigen patch test reactivity in patients with atopic dermatitis. *Clinical and Experimental Dermatology*, 1999, 29, 920-925.

[76] Guy-Grand, D; Vassalli P. Gut intraepithelial lymphocyte development. *Current Opinions in Immunology*, 2002, 14, 255-259.

[77] Hamid, Q; Boguniewicz, M; Leung, YM. Different in situ cytokine expression in acute versus chronic atopic dermatitis. *Journal of Clinical Investigation*, 1994, 94, 670-876.

[78] Hanifin, JM; Rajka, G. Diagnostic features of atopic dermatitis. *Acta Dermato-Venereologica (Stockholm)*, Suppl 1980, 92, 42-47.

[79] Hansen, TK; Host, A; Bindslev-Jensen, C. An evaluation of diagnostic value of different skin tests with egg in clinically egg-allergic children having atopic dermatitis. *Pediatric Allergy and Immunology*, 2004, 15, 428-434.

[80] Heine, RG; Verstege, A; Mehl, A; Staden, U; Rolinck-Werninghaus; Niggemann, B. Proposal for a standardized interpretation of the atopy patch test in children with atopic dermatitis and suspected food allergy. *Pediatric Allergy and Immunology*, 2006, 17, 213-217.

[81] Heinemann, C; Schliemann-Willers, S; Kelterer, D; Metzner, U; Kluge, K; Wiger-Alberti, W; Elsner, P. The atopy patch test – reproducibility and comparison of different evaluation methods. *Allergy*, 2002, 57, 641-645.

[82] Hernandez-Trujillo, VP; Nguyen, WT; Belleau, JT; Jeng, M; Conley, ME; Lew, DB. Cow's milk allergy in a patient with hyper-IgE syndrome- *Annals of Allergy, Asthma and Immunology*, 2004, 92, 469-474.

[83] Herrick, CA; MacLeod, H; Glusac, E; Tigelaar, RE; Bottomly, K. Th2 responses induced by epicutaneous or inhalational protein exposure are differently dependent on IL-4. *Journal of Clinical Investigation*, 2000, 105, 765-775.

[84] Holm, L; van Hage-Hamsten, M; Ohman, S; Scheynius, A. Sensitization to allergens of house-dust mite in adults with atopic dermatitis in a cold temperature region. *Allergy*, 1999, 54, 708-715.

[85] Holm, L; Matuseviciene, G; Scheynius, A; Tengvall Linder, M. Atopy patch test with house dust mite allergen: an IgE-mediated reaction? *Allergy*, 2004, 59, 874-882.

[86] Imayama, S; Hashizume, T; Miyahara, H; Tanahashi, T; Takeishi, M; Kubota, Y; Koga, T; Hori, Y; Fukuda, H. Combination of patch test and IgE for dust mite antigens differentiates L30 patients with atopic dermatitis into four groups. *Journal of American Academy of Dermatology*, 1992, 27, 531-538.

[87] Ingordo, V; D´Andria, C; Tortello Cannata, A. Reproducibility of the atopy patch test with whole house dust mite bodies in atopic dermatitis. *Contact Dermatitis*, 2000, 42, 174-175.

[88] Ingordo, V; D´Andria, G, D´Andria, C; Tortora, A. Result of atopy patch tests with house dust mites in adults with "intrinsic" and "extrinsic" atopic dermatitis. *Journal of European Academy of Dermatology and Venereology*, 2002, 16, 450-454.

[89] Ingordo, V; D´Andria, G; D´Andria, C. Adult-onset atopic dermatitis in a patch test population. *Dermatology*, 2003, 206, 197-203.

[90] Ingordo, V. The atopy patch test with whole dust mite bodies at 20%: Evaluation of out-comes in adult atopic dermatitis, non-eczematous atopic patients and healthy subjects. *Giornalle Italiano della Dermatologia e Venereologia*, 2004, 139, 195-206.

[91] Ingordo, V; Dalle Nogare, R; Colecchia, B; D´Andria, C. Is the atopy patch test with house dust mites specific for atopic dermatitis? *Dermatology*, 2004, 209, 276-283.

[92] Isolauri, E; Turjanmaa, K. Combined skin prick and patch testing enhances identification of food allergy in infants with atopic dermatitis. *Journal of Allergy and Clinical Immunology*, 1996, 97, 9-15.

[93] Järvinen, KM; Turpeinen, M; Suomalainen, H. Concurrent cereal allergy in children with cow's milk allergy manifested with atopic dermatitis. *Clinical and Exerimental* Allergy, 2003, 33, 1060-1066.

[94] Jesenak, M; Banovcin, P. Atopy patch test in the diagnosis of food allergy in children with atopic dermatitis. *Acta Medica (Hradec Kralove)*, 2006, 49, 199-201.

[95] Jesenak, M; Banovcin, P; Rennerova, Z; Havlicekova, Z; Jakusova, L; Ronchetti, R. Side effects of food atopy patch tests. *Clinical Pediatrics*, 2008 [Ahead of print].

[96] Johansson, SG; Bieber, T; Dahl, R; Friedmann, PS; Lanier, BQ; Lockey, RF; Motala, C; Ortega Martell, JA; Platts-Mills, TAE; Ring, J; Thien, F; Van Cauwenberge, P; Williams, HC. Revised nomenclature for allergy for global use: Report of the Nomenclature Review Committee of the World Allergy Organization, October 2003. *Journal of Allergy and Clinical Immunology*, 2001, 56, 813-824.

[97] Johansson, C; Eshaghi, H; Tengvall Linder, M; Jakobson, E; Scheynius, A. Positive atopy patch test reaction to *Malassezia furfur* in atopic dermatitis correlates with a T helper 2-like peripheral blood mononuclear response. *Journal of Investigational Dermatology*, 2002, 118, 1044-1051.

[98] Johansson, C; Sandstrom, MH; Bartosik, J; Sarnhult, T; Christianses, J; Zargari, A; Back, O; Wahlgren, CF; Faergemann, J; Scheynius, A; Tenqvall Linger M. Atopy patch test reactions to *Malassezia* allergens differentiate subgroup of atopic dermatitis patients. *British Journal of Dermatology*, 2003, 148, 479-488.

[99] Jonker, MJ; Bruynzeel, DP. The outcome of an additional patch-test reading on days 6 or 7. *Contact Dermatitis*, 2000, 42, 330-335.

[100] Kabelitz, D; Marischen, L; Oberg, HH; Holtmeier, W; Wesch, D. Epithelial defense by gamma T cells. *International Archives of Allergy and Immunology*, 2005, 137, 73-81.

[101] Kalach, N; Soulaines, P; de Boissieu, D; Dupont, C. A pilot study of the usefulness and safety of a ready-to-use atopy patch test (Diallertest) versus a comparator (Finn Chamber) during cow's milk allergy in children. *Journal of Allergy and Clinical Immunology*, 2005, 116, 1321-1326.

[102] Kaufman, HS; Roth, HL. Hyposensitization with alum precipitated extracts in atopic dermatitis: a placebo-controlled study. *Annals of Allergy*, 1974, 32, 321-330.

[103] Kekki, O; Turjanmaa, K; Isolauri, E. Differences in skin-prick and patch-test reactivity are related to the heterogeneity of atopic eczema in infants. *Allergy,* 1997, 52, 755-779.

[104] Kerschenlohr, K; Decard, S; Przybilla, B; Wollenberg, A. Atopy patch test reactions show a rapid influx of inflammatory dendritic epidermal cells in patients with extrinsic atopic dermatitis and patients with intrinsic atopic dermatitis. *Journal of Allergy and Clinical Immunology*, 2003, 111, 869-874.

[105] Kerschenlohr, K; Darsow, U; Burgdorf, WHC; Ring, J; Wollenberg, A. Lessons from atopy patch testing in atopic dermatitis. *Current Allergy and Asthma Reports*, 2004, 4, 285-289.

[106] Keskin, O; Tuncer, A; Adalioglu, G; Sekerel, BE; Sackesen, C; Kalayci, O. Evaluation of the utility of atopy patch testing, skin prick testing, and total and specific IgE assays in the diagnosis of cow's milk allergy. *Annals of Allergy, Asthma and Immunology*, 2005, 95, 553-560.

[107] Koehler, AM; Maibach, HI. Skin hyporeactivity in relation to patch testing. *Contact Dermatitis*, 2000, 42, 1-4.

[108] Krogulska, A; Wssowska-Krolikowska, K; Dynowski, J. Usefulness of atopy patch tests with food allergens in diagnosis of food allergy in children with dermatitis atopica. *Przeglad Pediatryczny,* 2007, 37, 245-249.

[109] Lack, G; Fox, D; Northstone, K; Golding, J. Factors associated with the development of peanut allergy in childhood. *New England Journal of Medicine*, 2003, 348, 977-985.

[110] Langeland, T; Braathen, LB; Borch, M. Studies of atopic patch tests. *Acta Dermato-Venereologica Suppl. (Stockholm)*, 1989, 144, 105-109.

[111] Langeveld-Wildschut, EG; van Marion, AM; Thepen, T; Mudde, GC; Bruijnzeel, PL; Bruijnzeel-Koomen, CA. Evaluation of variables influencing the outcome of the atopy patch test. *Journal of Allergy and Clinical Immunology*, 1995, 96, 66-73.

[112] Langeveld-Wildschut, EG; Thepen, T; Bihari, IC; ven Reijsen, FC; de Vries, IJ; Bruijnzeel, PL; Bruijnzeel-Koomen, CA. Evaluation of the atopy patch test and the cutaneous late-phase reaction as relevant models for the study of allergic inflammation in patients with atopic eczema. *Journal of Allergy and Clinical Immunology*, 1996, 98, 1019-1027.

[113] Langeveld-Wildschut, EG; Bruijnzeel, PLB; Mudde, GC; Versluis, C; Van Ieperen-Van Dijk, AG; Bihari, IC; Knol, EF; Thepen, T; Bruijnzeel-Koomen, CAFM; van Reijsen, FC. Clinical and immunologic variables in skin of patients with atopic eczema and either positive or negative atopy patch test reactions. *Journal of Allergy and Clinical Immunology*, 2000, 105, 1008-1016.

[114] Langeveld-Wildschut, EG; Riedl, H; Thepen, T; Bihari, IC; Bruijnzeel, PLB; Bruijnzeel-Koomen, CAFM. Modulation of the atopy patch test reaction by topical corticosteroids and tar. *Journal of Allergy and Clinical Immunology*, 2000, 106, 737-743.

[115] Lawrence, CM; Howel, D; Schuster, S. Site variation in anthralin inflammation on forearms skin. *British Journal of Dermatology*, 1986, 114, 609-613.

[116] Le Coz, CJ; Muller, B; Donnay, C. Marking patch test sites: description of a practical, clean, durable and inexpensive method. *Contact Dermatitis*, 2004, 49, 284-286.

[117] Leung, DY; Bieber, T. Atopic dermatitis. *Lancet*, 2003, 361, 151-160.

[118] Lindelof, B. A left versus right side comparative study of Finn Chamber™ patch tests in 220 consecutive patients. *Contact Dermatitis*, 1990, 22, 288-293.

[119] Lindelof, B. Regional variations of patch test response in nickel-sensitive patients. *Contact Dermatitis*, 1992, 26, 202.

[120] Magnusson, B; Hersle, K. Patch test methods (II). Regional variations of patch test responses. *Acta Dermato-Venereologica*, 1965, 45, 257-261.

[121] Maibach, H; Fregert, S; Magnusson, B; Pirila, V; Hjorth, N; Wilkinson, D; Malten, K; Lachapelle, JM; Calnan, C; Cronin, E. Quantification of the excited skin syndrome (the "angry back"). Retesting one patch at a time. *Contact Dermatitis*, 1982, 8, 78-83.

[122] Majamaa, H; Miettinen, A; Laine, S; Isolauri, E. Intestinal inflammation in children with atopic eczema: faecal eosinophil cationic protein and tumour necrosis factor-a as non-invasive indicators of food allergy. *Clinical and Experimental Allergy*, 1995, 26, 181-187.

[123] Majamaa, H; Moisio, P; Holm, K, Kautiainen, H; Turjanmaa, K. Cow's milk allergy, diagnostic accuracy of skin prick and patch tests and specific IgE. *Allergy*, 1999, 54, 346-351.

[124] Majamaa, H; Moisio, P; Holm, K; Turjanmaa, K. Wheat allergy: diagnostic accuracy of skin prick and patch tests and specific IgE. *Allergy*, 1999, 54, 851-856.

[125] Manzini, BM; Motolese, A; Donini, M; Seidenari, S. Contact allergy to *Dermatophagoides* in atopic dermatitis patients and healthy subjects. *Contact Dermatitis*, 1995, 33, 243-246.

[126] Marsella, R; Olivry, T; Maeda, S. Cellular and cytokine kinetics after epicutaneous allergen challenge (atopy patch testing) with house dust mites in high-IgE beagles. *Veterinary Dermatology*, 2006, 17, 111-120.

[127] Matzinger, M; Daneman, A. Esophageal involvement in eosinophilic gastroenteritis. *Pediatric Radiology*, 1983, 13, 35-38.

[128] Maurer, D; Ebner, C; Reininger, B; et al. The high affinity IgE receptor ($Fc_{\varepsilon}RI$) mediates IgE-dependent allergen presentation. *Journal of Immunology*, 1995, 154, 6285-6290.

[129] Meglio, P; Milita, O; Businco, L. Patch test response to house dust mites is positive in children with atopic dermatitis and their parents. *Journal of Investigational Allergology and Clinical Immunology*, 1996, 6, 190-195.

[130] Mehl, A; Rolinck-Werninghaus, C; Staden, U; Verstege, A; Beyer, K; Niggemann, B. The atopy patch test in the diagnosis of suspected food-related symptoms in children. *Journal of Allergy and Clinical Immunology*, 2006, 118, 923-929.

[131] Memon, AA; Friedmann, PS. Studies on the reproducibility of allergic contact dermatitis. *British Journal of Dermatology*, 1996, 134, 208-214.

[132] Mitchell, JC. The angry back syndrome: eczema creates eczema. *Contact Dermatitis*, 1975, 1, 193-197.

[133] Mitchell, EB; Crow, J; Chapman, MD; Jouhal, SS; Pope, FM; Platts-Mills, TAE. Basophils in allergen-induced patch test sites in atopic dermatitis. *Lancet*, 1982, 1: 127-130.

[134] Mitchell, J; Maibach, HI. Managing the excited skin syndrome: patch testing hyperirritable skin. *Contact Dermatitis*, 1997, 37, 193-199.

[135] Motomura, C; Shibata, R; Nashie, H. The atopy patch test for food allergy in children with atopic dermatitis. *Allergy in Practice*, 2001, 277, 565-569.

[136] Moyer, P. New Patch Test Increases Accuracy, Convenience in Diagnosing Cow's Milk Allergy. *American Academy of Allergy, Asthma and Immunology 61st Annual Meeting*: Abstract 247, 2005.

[137] Nedorost, ST; Cooper, KD. The role of patch testing for chemical and protein allergens in atopic dermatitis. *Current Allergy and Asthma Report*, 2001, 1, 323-328.

[138] Nielsen, RG; Binslev-Jensen, C; Kruse-Andersen, S; Husby, S. Severe gastroesophageal reflux disease and cow milk hypersensitivity in infants and children: disease association and evaluation of a new challenge procedure. *Gastroenterology and Nutrition*, 2004, 93, 383-391.

[139] Niggemann, B; Sielaff, B; Beyer, K; Binder, C; Wahn, U. Outcome of double-blind, placebo-controlled food challenge tests in 107 children with atopic dermatitis. *Clinical and Experimental Allergy*, 1999, 29, 91-96.

[140] Niggemann, B; Reibel, S; Wahn, U. The atopy patch test (APT) – a useful tool for the diagnosis of food allergy in children with atopic dermatitis. *Allergy*, 2000, 55, 281-285.

[141] Niggemann, B. The role of the atopy patch test (APT) in diagnosis of food allergy in infants and children with atopic dermatitis. *Pediatric Allergy and Immunology*, 2001, 12 (Suppl 14), 37-40.

[142] Niggemann, B; Reibel, S; Roehr, CC; Felger, D; Ziegert, M; Sommerfeld, C; Wahn, U. Predictors of positive food challenge outcome in non-IgE-mediated reactions to food in children with atopic dermatitis. *Journal of Allergy and Clinical Immunology*, 2001, 108, 1053-1058.

[143] Niggemann, B. Evolving role of the atopy patch test in the diagnosis of food allergy. *Current Opinions in Allergy and Clinical Immunology*, 2002, 2, 253-256.

[144] Niggemann, B; Zeigert, M; Reibel, S. Importance of chamber size for the outcome of atopy patch testing in children with atopic dermatitis and food allergy. *Journal of Allergy and Clinical Immunology*, 2002, 110, 515-516.

[145] Niggemann, B. Atopy patch test (APT) – its role in diagnosis of food allergy in atopic dermatitis. *Indian Journal of Pediatrics*, 2002, 69, 57-59.

[146] Niggemann, B. Evolving role of the atopy patch test in the diagnosis of food allergy. *Current Opinions in Allergy and Clinical Immunology*, 2002, 2, 253-256.

[147] Niggemann, B; Beyer, K. Diagnostic pitfalls in food allergy in children. *Allergy*, 2005, 60, 104-107.

[148] Norgaard, A; Skov, PS; Bindslev-Jensen, C. Egg and milk allergy in adults: comparison between fresh foods and commercial allergen extracts in skin prick test and histamine release from basophils. *Clinical and Experimental Allergy*, 1992, 22, 940-947.

[149] Norris, P; Schofield, O; Camp, R. A study of the role of house dust mite in atopic dermatitis. *British Journal of Dermatology*, 1988, 118, 435-440.

[150] Novak, N; Haberstok, J; Geiger, E; Bieber, T. Dendritic cells in allergy. *Allergy*, 1999, 55, 792-803.

[151] Nowak-Wegrzyn, A; Conover-Walekr; MK; Wood, RA. Food-allergic reactions in schools and preschools. *Archives of Pediatric and Adolescent Medicine*, 2001, 155, 790-795.

[152] O'Leary, PF; Shanahan, F. Food allergies. *Current Gastroenterology Reports*, 2002, 4, 373-382.

[153] Oldhoff, JM; Bihari, IC; Knol, EF; Bruijnzeel-Koomen, CA; de Bruin-Weller, MS. Atopy patch test in patients with atopic eczema/dermatitis syndrome: comparison of petrolatum and aqueous solution as a vehicle. *Allergy*, 2004, 59, 451-456.

[154] Oldhoff, JM; Knol, EF; Laaper-Ertmann, M; Bruijnzeel-Koomen, CAFM; de Bruin-Weller, MS. Modulation of the atopy patch test: tacrolimus 0.1% compared with triamcinolone acetonide 0.1%. *Allergy*, 2006, 61, 622-628.

[155] Oldhoff, JM; Darsow, U; Werfel, T; Bihari, IC; Katzer, K; Laifaoui, J; Plotz, S; Kapp, A; Knol, EF; Bruijnzeel-Koomen, CAFM; Ring, J; de Bruin-Weller, MS. No effect of anti-interleukin-5 therapy (mepolizumab) on the atopy patch test in atopic dermatitis patients. *International Archives of Allergy and Immunology*, 2006, 141, 290-294.

[156] Olivry, T; Deangelo, KB; Dunston, SM; Clarke, KB; Mccall, CA. Patch testing of experimentally sensitized beagle dogs: development of a model for skin lesions of atopic dermatitis. *Veterinary Dermatology*, 2006, 17, 95-102.

[157] Oranje, AP; Van Gysel, D; Mulder, PGH; Dieges, PH. Food-induced contact urticaria syndrome (CUS) in atopic dermatitis: reproducibility of repeated and duplicated testing with a skin provocation test, the skin application food test (SAFT). *Contact Dermatitis*, 1994, 31, 314-318.

[158] Osterballe, M; Andersen, KE; Bindslev-Jensen, C. The diagnostic accuracy of the atopy patch test in diagnosing hypersensitivity to cow's milk and hen's egg in unselected

children with and without atopic dermatitis. *Journal of American Academy of Dermatology*, 2004, 51, 556-562.

[159] Partidos, CD; Beignon, AS; Brown, F; Kramer, E; Briand, JP, Muller, S. Applying peptide antigens onto bare skin: induction of humoral and cellular responses and potential for vaccination. *Journal of Controlled Release*, 2002, 85, 27-34.

[160] Pastorello, EA. Skin tests for diagnosis of IgE-mediated allergy. *Allergy*, 1993, 48, 57-62.

[161] Perackis, K; Celik-Bilgili, S; Staden, U; Mehl, A; Niggemann, B. Influence of age on the outcome of the atopy patch test with food in children with atopic dermatitis. *Journal of Allergy and Clinical Immunology*, 2003, 112, 625-627.

[162] Pickler, LJ; Treer, JR; Ferguson-Darnell, B; Collins, PA; Bergstresser, PR; Terstappen, LW. Control of lymphocyte recirculation in man II. Differential regulation of the cutaneous lymphocyte-associated antigen, a tissue-selective homing receptor for skin-homing T cells. *Journal of Immunology*, 1993, 150, 1122-1136.

[163] Pigatto, PD; Bigardi, AS; Valsecchi, RH; Di Landro, A. Mite patch testing in atopic eczema: a search for correct concentration. *Australian Journal of Dermatology*, 1997, 38, 231-232.

[164] Przybilla, B; Ring, J. Food allergy and atopic eczema. *Seminars in Dermatology*, 1990, 9, 220-225.

[165] Rajka, G. Essential aspects of atopic dermatitis. Berlin: Springer, 1989.

[166] Rancé, F. What is the optimal occlusion time for the atopy patch test in the diagnosis of food allergies in children with atopic dermatitis? *Pediatric Allergy and Immunology*, 2004, 15, 93-96.

[167] Räsänen, L; Lehto, M; Turjanmaa, K; Savolainen, J; Reunala, T. Allergy to ingested cereals in atopic children. *Allergy*, 1994, 49, 871-876.

[168] Reitamo, S; Visa, K; Kahonen, K; Kayhko, K; Stubb, S; Salo, OP. Eczematous reactions in atopic patients caused by epicutaneous testing with inhalant allergens. *British Journal of Dermatology*, 1986, 114, 303-309.

[169] Remón, ZL; Lebrero, EA; Fernández, EM; Molero, MIM. Food protein-induced enterocolitis syndrome caused by fish. *Allergologia et Immunopathologia*, 2005, 33, 312-316

[170] Ring, J; Kunz, B; Bieber, T; Vieluf, D; Przybilla, B. The "atopy patch test" with aeroallergens in atopic eczema. *Journal of Allergy and Clinical Immunology*, 1989, 82, 195.

[171] Ring, J; Darsow, U; Abeck, D. The atopy patch test as a method of studying aeroallergens as triggering factors of atopic eczema. *Dermatologic Treatment*, 1996, 1, 51-60.

[172] Roehr, CC; Reibel, S; Ziegert, M; Sommerfeld, C; Wahn, U; Niggemann, B. Atopy patch test together with level of specific IgE reduces the need for oral food challenge in children with atopic dermatitis. *Journal of Allergy and Clinical Immunology*, 2001, 107, 548-553.

[173] Rokaité, R; Labanauskas, L; Vaideliené, L. Role of the skin patch test in diagnosing food allergy in children with atopic dermatitis. *Medicina (Kaunas)*, 2004, 4, 1081-1087.

[174] Ronchetti, R; Jesenak, M; Barreto, M; Trubacova, D; Rennerova, Z; Pohanka, V; Banovcin, P; Villa, MP. The epidemiology of food atopy patch tests: a study of 355

unselected school children aged 10 from two European countries. *Acta Medica Martiniana*, 2007, 7, 16-25.

[175] Ronchetti, R; Jesenak, M; Trubacova, D; Pohanka, V; Villa, MP. Epidemiology of atopy patch tests with food and inhalant allergens in an unselected population of children. *Pediatric Allergy and immunology*, 2008, doi:10.1111/j.1399-3038.2007.00712.x.

[176] Ronchetti, R; Jesenak, M; Barberi, S; Ronchetti, F; Rennerova, Z; Trubacova, D; Villa, MP. Reproducibility of atopy patch tests with food and inhalant allergens. *Journal of Biological Regulators and Homeostatic Agents*, 2008, 22, 27-33.

[177] Rostenberg, A; Sulzberger, MD. Some results of patch tests. *Archives of Dermatology*, 1937, 35, 433-454.

[178] Rothman, S. Physiology and biochemistry of the skin. Chicago: University of Chicago Press, 1954, 120-152.

[179] Ruchelli, E; Wenner, W; Voytek, T; Brown, K; Liacouras. Severity of esophageal eosinophilia predicts response to conventional gastroesophageal reflux therapy. *Pediatric and Developmental Pathology*, 1999, 2, 15-18.

[180] Saarinen, KM; Suomalainen, H; Savilahti, E. Diagnostic value of skin-prick and patch tests and serum eosinophil cationic protein and cow's milk-specific IgE in infants with cow's milk allergy. *Clinical and Experimental Allergy*, 2001, 31, 423-429.

[181] Sager, N; Feldmann, A; Schilling, G; Kreitsch, P; Neumann, C. House dust mite-specific T cells in the skin of subjects with atopic dermatitis: frequency and lymphokine profile in the allergen patch test. *Journal of Allergy and Clinical Immunology*, 1992, 89, 801-810.

[182] Saloga, J; Knop, J. Does sensitization through the skin occur? *Allergy*, 2000, 55, 905-909.

[183] Samochocki, Z; Owczarek, W; Zabielski, S. Can atopy patch tests aeroallergens be an additional diagnostic criterion for atopic dermatitis. *European Journal of Dermatology*, 2006, 16, 151-154.

[184] Sampson, HA. Immunologically mediated food allergy: the importance of food challenge procedures. *Annals of Allergy*, 1988, 60, 646-651.

[185] Sampson, HA; Ho DG. Relationship between food-specific IgE concentrations and the risk of positive food challenge in children and adolescents. *Journal of Allergy and Clinical Immunology*, 1997, 100, 444-451.

[186] Sampson, HA. Food allergy. Part 2: Diagnosis, management. *Journal of Allergy and Clinical Immunology*, 1999, 103, 981-989.

[187] Sampson, HA. The evaluation and management of food allergy in atopic dermatitis. *Clinics in Dermatology*, 2003, 21, 183-192.

[188] Sampson, HA. Update on food allergy. *Journal of Allergy and Clinical Immunology*, 2004, 113, 805-819.

[189] Schiessl, C; Wolber, C; Strohal, R. Reproducibility of patch tests: comparison of identical test allergens from different commercial sources. *Contact Dermatitis*, 2004, 50, 27-33.

[190] Schleimer, RP; Togias, AG. Introduction: systemic aspects of allergic disease. *Journal of Allergy and Clinical Immunology*, 2000, 106 Pt 2, 191.

[191] Seidenari, S; Manzini, BM; Danese, P; Di Nardo, A; Giannetti, A. Modified patch test with whole mite cultures: experience on 31 subjects affected by atopic dermatitis. *Giornalle Italiano della Dermatologia e Venereologia*, 1991, 126, 5-10.

[192] Seidenari, S; Manzini, BM; Danese, P; Gianetti, A. Positive patch tests to whole house dust mite culture and purified mite extracts in patients with atopic dermatitis, asthma and rhinitis. *Annals of Allergy*, 1992, 69, 201-205.

[193] Seidenari, S; Di Nardo A. Cutaneous reactivity to allergens at 24-h increases from the antecubital fossa to the wrist: an echografic evaluation by means of a new image analysis system. *Contact Dermatitis*, 1992, 26, 171-176.

[194] Seidenari, S; Giusti, F; Bertoni, L. Reproducibility of APT. *Allergy*, 2002, 57, 1082.

[195] Seidenari, S; Giusti, F; Bertoni, L; Mantovani, L. Combined skin prick and patch testing enhances identification of peanut-allergic patients with atopic dermatitis. *Allergy*, 2003, 58, 495-499.

[196] Seidenari, S; Giusti, F; Pellacani, G; Bertoni; L. Frequency and intensity of responses to mite patch tests are lower in non atopic subjects in respect to patients with atopic dermatitis. *Allergy*, 2003, 58, 426-429.

[197] Shaw, DW; Zhai, H; Maibach, HI; Niklasson, B. Dosage consideration in patch testing with liquid allergens. *Contact Dermatitis*, 2002, 47, 86-90.

[198] Sicherer, SH; Munoz-Furlong, A; Murphy, R; et al. Symposium: Pediatric food allergy. *Pediatrics*, 2004, 111, 1591-1594.

[199] Silva Segundo, GR. Food allergy and atopy patch tests. *Jornal de Pediatria (Rio J)*, 2007, 83, 381-382.

[200] Simonetti, V; Manzini, BM; Seidenari, S. Patch testing with nickel sulfate: comparison between 2 nickel sulfate preparations and 2 different test sites on the back. *Contact Dermatitis*, 1998, 39, 187-191.

[201] Sinagra, JL; Bordignon, V; Ferraro, C; Cristaudo, A; Di Rocco, M; Amorosi, B; Capitanio, B. Unnecessary milk elimination diets in children with atopic dermatitis. *Pediatric Dermatology*, 2007, 24, 1-6.

[202] Soury, D; Barratt, G; Ah-Leung, S; Legrand, S; Chacun, H; Ponchel, G. Skin localization of cow's milk proteins delivered by a new ready-to-use atopy patch test. *Pharmaceutical Research*, 2005, 22, 1530 1536.

[203] Spergel, JM; Mizoguchi, E; Brewer, JP; Martin, TR; Bhan, Ak; Geha, RS. Epicutaneous sensitization with protein antigen induces localized allergic dermatitis and hyperresponsiveness to methacholine after single exposure to aerosolized antigen in mice. *Journal of Clinical Investigation*, 1998, 101, 1614-1622.

[204] Spergel, JM; Beausoleil, JL; Mascarenhas, M; Liacouras, CA. The use of skin prick tests and patch tests to identify causative foods in eosinophilic esophagitis. *Journal of Allergy and Clinical Immunology*, 2002, 109, 363-368.

[205] Spergel, JM; Andrews, T; Brown-Whitehorn, TF; Beausoleil, JL; Liacouras, CA. Treatment of eosinophilic esophagitis with specific food elimination diet directed by a combination of skin prick and patch tests. *Annals of Allergy, Asthma and Immunology*, 2005, 95, 336-343.

[206] Spergel, JM; Brown-Whitehorn, T. The use of patch testing in the diagnosis of food allergy. *Current Allergy and Asthma Reports*, 2005, 5, 86-90.

[207] Spergel, JM; Beausoleil, JL; Brown-Whitehorn, T; Liacouras, CA. Authors′ response to detection of causative foods by skin prick and atopy patch tests in patients with

eosinophilic esophagitis: things are not what they seem. *Annals of Allergy, Asthma and Immunology*, 2006, 96, 367-378.

[208] Spergel, JM; Brown-Whitehorn, T; Beausoleil, JL; Shuker, M; Liacouras, CA. Predictive values for skin prick test and atopy patch test for eosinophilic esophagitis. *Journal of Allergy and Clinical Immunology*, 2007, 119, 509-511.

[209] Strid, J; Hourihane, J; Kimber, I; Callard, R; Strobel, S. Disruption of the stratum corneum allows potent epicutaneous immunization with protein antigens resulting in a dominant systemic Th2 response. *European Journal of Immunology*, 2004, 34, 2100-2109.

[210] Strid, J; Hourihane, J; Kimber, I; Callard, R; Strobel, S. Epicutaneous exposure to peanut protein prevents oral tolerance and enhances allergic sensitization. *Clinical and Experimental Allergy*, 2005, 35, 757-766.

[211] Stromberg, L. Diagnostic accuracy of the atopy patch test and skin-prick test for the diagnosis of food allergy in young children with atopic eczema/dermatitis syndrome. *Acta Paediatrica*, 2002, 91, 1044-1049.

[212] Tagami, H. Measurement of electrical conductance and impedance. In: Serup, J; Jemec, GBE, editors. Handbook of non-invasive methods and the skin. Boca Raton: CRC Press, 1995, 159-164.

[213] Tanaka, Y; Anan, S; Yoshida, H. Immunohistochemical studies in mite antigen-induced patch test sites in atopic dermatitis. *Journal of Dermatologic Sciences*, 1990, 1, 361-368.

[214] Taskapan, O; Digan, B; Harmanyeri, Y. Atopy patch test reactivity to house dust mites in patients with scabies. *International Journal of Dermatology*, 2003, 42, 244-248.

[215] Teuber, S; Porch-Curren, C. Unproved diagnostic and therapeutic approaches to food allergy and intolerance. *Current Opinions in Allergy and Clinical Immunology*, 2003, 3, 217-221.

[216] Thappa, DM. Relevant investigations in atopic dermatitis. *Indian Journal for the Practicing Doctor*, 2006, 3, 11-12.

[217] Thepen, T; Langeveld-Wildschut, EG; Bihari, IC; van Wlchen, DF; van Reijsen, FC; Mudde, GC; Bruijnzeel-Koomen, CAFM. Biphasic response against aeroallergen in atopic dermatitis showing a switch from an initial TH2 response to a TH1 response in situ: an immunohistochemical study. *Journal of Allergy and Clinical Immunology*, 1996, 97, 828-837.

[218] Turjanmaa, K. "Atopy patch tests" in the diagnosis of delayed food hypersensitivity. *European Annals of Allergy and Clinical Immunology*, 2002, 34, 363-367.

[219] Turjanmaa, K. The role of atopy patch tests in the diagnosis of allergy in atopic dermatitis. *Current Opinion in Allergy and Clinical Immunology*, 2005, 5, 425-428.

[220] Turjanmaa, K; Darsow, U; Niggemann, B; Rancé, F; Vanto, T; Werfel, T. EAACI/GA^{2}LEN Position paper: Present status of the atopy patch test. *Allergy*, 2006, 61, 1377-1384.

[221] Uehara, M; Ofuji, S. patch test reactions to human dander in atopic dermatitis. *Archives of Dermatology*, 1976, 112, 951-954.

[222] van Bever, HP; Doex, M; Stevens, WJ. Food and food additives in severe atopic dermatitis. *Allergy*, 1989, 44, 588-594.

[223] van Reijsen, FC; Bruijnzeel-Koomen, CAFM; Kalthoff, FS. Skin-derived aeroallergen-specific T-cell clones of Th2 phenotype in patients with atopic dermatitis. *Journal of Allergy and Clinical Immunology*, 1992, 90, 184-192.

[224] van der Valk, PGM; Maibach, HI. Potential for irritation increases from the wrist to the cubital fossa. *British Journal of Dermatology*, 1985, 121, 709-712.

[225] van der Valk, PGM; Kruis-de Vries, MH; Nater, JP; Bleumink, E; de Jong, MCJM. Eczematous (irritant and allergic) reactions of the skin and barrier function as determined by water vapour loss. *Clinical and Experimental Dermatology*, 1985, 10, 185-193.

[226] van Reijsen, FC; Bruijnzeel-Koomen, CA; Kalthoff, FS; Maggi, E; Romagnani, S; Westland, JK; Mudde, GC. Skin-derived aeroallergen-specific T-cells clones of Th2 phenotype in patients with atopic dermatitis. *Journal of Allergy and Clinical Immunology*, 1992, 90, 184-193.

[227] van Strien, GA; Korstanje, MJ. Site variations in patch test responses on the back. *Contact Dermatitis*, 1994, 31, 95-96.

[228] van Voorst Vader, PC; Lier, JG; Woest, TE; Coenraads, PJ; Nater, JP. Patch tests with house dust mite antigens in atopic dermatitis patients: methodological problems. *Acta Dermato-Venereologica (Stockholm)*, 1991, 71, 301-305.

[229] Vaali, K; Puumalainen, TJ; Lehto, M; Wolff, H; Rita, H; Alenius, H; Palosuo, T. Murine model of food allergy after epicutaneous sensitization: role of mucosal mast cell protease-1. *Scandinavian Journal of Gastroenterology*, 2006, 41, 1405-1413.

[230] Vanto, T; Juntunen-Backman, K; Kalimo, K; Klemola, T; Koivikko, A; Koskinen, J; Syvänen, P; Valovirta, E; Varjonen, E. The patch test, skin prick test, and serum milk-specific IgE as diagnostic tools in cow's milk allergy in infants. *Allergy*, 1000, 54, 837-842.

[231] Varela, P; Selores, M; Gomes, E; Silva, E; Matos, E; dos Santos, L; Amado, J; Massa, A. Immediate and delayed hypersensitivity to mite antigens in atopic dermatitis. *Pediatric Dermatology*, 1999, 16, 1-5.

[232] Ventura, A; Salvatore, CM; Longo, G. Might laboratory tests other than RAST help in diagnosing food allergy? *Pediatric Allergy and Immunology*, 1995, 6 Suppl 8, 44-48.

[233] Veremis-Ley, M; Ramirez, H; Baron, E; Hanneman, K; Lankerani, L; Scull, H; Cooper, KD; Nedorost, ST. Laser-assisted penetration of allergens for patch testing. *Dermatitis*, 2006, 17, 15-22.

[234] Wang, LF; Lin, JY; Hsieh, KH; Lin, RH. Epicutaneous exposure of protein antigens induces a predominant Th2-like response with high IgE production in mice. *Journal of Immunology*, 1996, 156, 4077-4082.

[235] Weiner, HL. Oral tolerance: immune mechanisms and treatment of autoimmune diseases. *Immunology Today*, 1997, 18, 335-343.

[236] Weissenbacher, S; Bacon, T; Targett, D; Behrendt, H; Ring, J; Darsow, U. Atopy patch test – reproducibility and elicitation of itch in different application sites. *Acta Dermato-Venereologica (Stockholm)*, 2005, 85, 147-151.

[237] Weissenbacher, S; Traidl-Hoffman, C; Eyerich, K; Katzer, K; Braeutigam, M; Loeffler, H; Hofmann, H; Behrendt, H; Ring, J; Darsow, U. Modulation of atopy patch test and skin prick test by pretreatment with 1% pimecrolimus cream. *International Archives of Allergy and Immunology*, 2006, 140, 239-244.

[238] Werfel, T; Kapp, A. Environmental and other major provocation factors in atopic dermatitis. *Allergy*, 1998, 53, 731-739.

[239] Werfel, T; Ahlers, G; Schmidt, P; Boecker, M; Kapp, A; Neumann, C. Milk-responsive atopic dermatitis is associated with a casein-specific lymphocyte response in adolescent and adult patients. *Journal of Allergy and Clinical Immunology*, 1997, 99, 124-133.

[240] Whitmore, SE; Sherertz, EF; Belsito, DV; Maibach, HI; Nethercott, JR. Aeroallergen patch testing for patients presenting to contact dermatitis clinics. *Journal of American Academy of Dermatology*, 1996, 35, 700-704.

[241] Wistokat-Wűlfing, A; Schmidt, P; Darsow, U; Ring, J; Kapp, A; Werfel, T. Atopy patch test reactions are associated with T lymphocyte-mediated allergen-specific immune responses in atopic dermatitis. *Clinical and Experimental Allergy*, 1999, 29, 513-521.

[242] Wollenberg, A; de la Salle, H; Hanau, D; Liu, FT; Bieber, T. Human keratinocytes release the endogenous β-galactoside-bonding soluble lectin εBP which binds to langerhans cells where it modulates their binding capacity for IgE glycoforms. *Journal of Experimental Medicine*, 1993, 178, 777-785.

[243] Wollenberg, A; Kraft, S; Hanau, D; Bieber, T. Immunomorphological and ultrastructural characterization of Langerhans cells and a novel, inflammatory dendritic epidermal cell (IDEC) population in lesional skin of atopic eczema. *Journal of Investigative Dermatology*, 1996, 106, 446-453.

[244] Wűthrich, B. Zur Nahrungsmittelallergie. *Allergologie*, 1993, 16, 280-287.

[245] Young, E; Bruijnzeel-Koomen, CA; Berrens, L. Delayed type hypersensitivity in atopic dermatitis. *Acta Dermatovenereologica Suppl.*, 1982, 114, 77-81.

[246] Zeiger, RS. Atopy in infancy and early childhood: Natural history and role of skin testing. *Journal of Immunology and Clinical Allergy*, 1985, 75, 633-639.

[247] Zuang, V; Archer, G; Rona, C; Vignini, M; Mosca, M; Berardesca, E. Predicting visual assessment of allergic patch test reactions by non-invasive measurements. *Skin Pharmacology and Applied Skin Physiology*, 2000, 13, 39-51.

In: Food Allergies: New Research
Editor: Carrie M. Chesterton, pp. 91-120
ISBN 978-1-60456-978-0

Chapter 2

CLINICAL MANIFESTATIONS OF FOOD ALLERGY

Alexander K.C. Leung[1,*] and Deepak Kamat[2]
[1] University of Calgary, Alberta Children's Hospital,
#200, 233 – 16th Avenue NW, Calgary, Alberta, T2M 0H5, Canada
[2] Carman and Ann Adams Department of Pediatrics, Wayne State University,
Children's Hospital of Michigan, Detroit, Michigan, USA

ABSTRACT

Food allergy is defined as an adverse reaction because of an abnormal immunological response to food protein. The immune pathogenesis is, in the majority of cases, IgE-mediated although it may also be cell-mediated (non-IgE) or mixed IgE/cell-mediated. Food allergy affects as many as 2 to 8% of young children and the presentation can be highly variable. There is usually a clear temporal relationship between food exposure and the development of allergic symptoms. At times, symptoms may develop hours or days after food exposure making the diagnosis difficult.

Food allergy usually presents as multi-system involvement, most commonly gastrointestinal symptoms which occur with a frequency of 50 to 80% of cases. These are followed by cutaneous symptoms and respiratory symptoms, occurring in 20 to 40%, and 4 to 25% of cases, respectively. Gastrointestinal manifestations include oral allergy syndrome, gastrointestinal anaphylaxis, allergic eosinophilic esophagitis, allergic eosinophilic gastroenteropathy, food protein-induced enteropathy, food protein-induced enterocolitis syndrome, food protein-induced proctocolitis, gluten-sensitive enteropathy, infantile colic, irritable bowel syndrome, and constipation. Cutaneous manifestations are urticaria/angioedema, atopic dermatitis, contact dermatitis, and dermatitis herpetiformis. Rhinitis/rhinoconjunctivitis, asthma, Heiner syndrome, and serous otitis media are the respiratory manifestations of food allergy. Other manifestations include systemic anaphylaxis, food-dependent exercise-induced anaphylaxis, migraine, epilepsy, diabetes mellitus, nephrotic syndrome, nocturnal enuresis, anemia, thrombocytopenia, vasculitis, and arthropathy/arthritis. This chapter discusses the various clinical manifestations of food allergy.

---

* Correspondence to: Dr. Alexander K.C. Leung, #200, 233 – 16th Avenue NW, Calgary, Alberta, T2M 0H5 Canada Telefax: (403) 230-3322; e-mail: aleung@ucalgary.ca

Keywords: food allergy, clinical manifestations

INTRODUCTION

Food allergy is defined as an adverse reaction because of an abnormal immunological response to a food or food additive [1,2]. The reaction occurs only in some patients, may occur after a small amount of the offending substance being ingested, and is unrelated to any physiological effect of the food or food additive [1,2]. The immune pathogenesis is, in the majority of cases, IgE-mediated although it may also be cell-mediated (non-IgE) or mixed IgE/cell-mediated. IgE-mediated reactions are caused by inflammatory mediators and cytokines released when circulating food allergens bind to specific IgE residing on the surface of mast cells and basophils. These reactions are associated with rapid development of symptoms, usually within minutes to 2 hours while cell-mediated reactions develop over hours or days [3,4]. Mixed IgE/cell-mediated reactions are characterized by intense eosinophilic infiltration of the specific organ involved and may lead to chronic disorders such as atopic dermatitis and allergic eosinophilic gastroenteropathy [3,5]. Food allergy affects approximately 4 to 8% of young children and 1 to 4% of adults [1,3,6-9]. The presentations of food allergy can be highly variable. There is usually a clear temporal relationship between food exposure and the development of allergic symptoms [10]. At times, symptoms may develop hours or even days after food exposure making the diagnosis difficult.

Table 1. Clinical Manifestations of Food Allergy

Gastrointestinal manifestations
Oral allergy syndrome Gastrointestinal anaphylaxis Allergic eosinophilic esophagitis Allergic eosinophilic gastroenteropathy Food protein-induced enteropathy Food protein-induced enterocolitis syndrome Food protein-induced proctocolitis Gluten-sensitive enteropathy Infantile colic Irritable bowel syndrome Recurrent abdominal pain Constipation
Cutaneous manifestations
Urticaria/angioedema Atopic dermatitis Contact dermatitis Dermatitis herpetiformis

Table 1. (Continued)

Respiratory manifestations
Rhinitis/rhinoconjunctivitis Chronic sinusitis Asthma Heiner syndrome Serous otitis media Ménière's disease
Generalized manifestations
Systemic anaphylaxis Food-dependent exercise-induced anaphylaxis Hyperactivity
Neurologic manifestations
Migraine Epilepsy
Endocrine manifestation
Diabetes mellitus
Renal manifestations
Nephrotic syndrome Nocturnal enuresis
Hematologic manifestations
Anemia Thrombocytopenia
Cardiovascular manifestation
Vasculitis
Rheumatic manifestation
Arthropathy/arthritis

The clinical manifestations of food allergy are listed in Table 1. Food allergy usually presents as multi-system involvement, most commonly gastrointestinal symptoms with a frequency between 50% and 80%, followed by cutaneous symptoms (in 20 to 40%) and respiratory symptoms (in 4 to 25%) [11,12]. Symptoms may be mild or severe and most often occur within one to two hours after the offending food has been eaten. Occasionally, the onset of symptoms may be delayed for 48 to72 hours. From a clinical and diagnostic stand-point, it is most useful to subdivide the clinical manifestations according to the predominant organ or system of involvement. This chapter discusses the various clinical manifestations of food allergy. It is important to remember that the book was written in accordance to the patient but the patient does not always get sick in accordance to the book.

Gastrointestinal Manifestations

Oral Allergy Syndrome

Oral allergy syndrome (pollen-food allergy syndrome) is a complex of symptoms induced by exposure of the oral and pharyngeal mucosa to plant protein allergens [13,14]. Patients are usually sensitized to an aeroallergen initially [15]. The IgE antibodies to the aeroallergen cause the oral allergy syndrome [16]. Botanical cross-reactivity as a result of shared epitopes between pollen and causative fruits and vegetables has been suggested as a possible mechanism of local mast cell activation [17]. The oral allergy syndrome is considered a form of contact urticaria that is confined mainly to the oropharynx [2]. Symptoms include rapid onset of itching, tingling, burning, and/or angioedema of the lips, tongue, palate, and throat within minutes of ingestion of fresh fruit and vegetable [14]. Symptoms usually resolve rapidly. Occasionally, the clinical course is more dramatic with potentially fatal pharyngeal swelling or progression towards a generalized anaphylactic reaction [10,18]. The syndrome generally occurs in patients with inhalant allergy to birch, mugwort, or ragweed pollen and is associated with the ingestion of various fresh fruits (e.g., bananas, melons, citrus fruits) and raw vegetables (e.g. carrots, tomatoes, celery) [19-22]. It is uncommon to have several fruits and vegetables that cause the oral allergy syndrome in one patient [23]. However, allergy to ragweed may cross-react when exposed to fresh melon and banana [24]. Oral allergy syndrome is more prevalent in adults than in children [25]. Most patients have some degree of allergic conjunctivitis or allergic rhinitis because the IgE antibodies to an aeroallergen cross-react with the fruit or vegetable proteins [15,26]. It is interesting to note that if the offending fruit or vegetable is cooked, then the patient does not usually experience any symptom as the food allergens are generally destroyed by heating [4,15]. Patients who remain sensitive to cooked fruit or vegetable may be sensitive to proteins that do not cross-react with pollens and do not actually have oral allergy syndrome [21,27]. Although these patients react to food typically associated with oral allergy syndrome, the absence of pollenosis and presence of symptoms beyond the oropharynx suggest conventional food allergy rather than oral allergy syndrome [21].

Gastrointestinal Anaphylaxis

Gastrointestinal anaphylaxis is an IgE-mediated gastrointestinal hypersensitivity that often accompanies other systemic manifestations of food allergy [28,29]. This may be manifested as nausea, vomiting, abdominal pain, flatulence, abdominal distension, or diarrhea [2]. The reaction usually occurs within minutes to 2 hours of food ingestion [30]. Repeated ingestion of a food allergen may induce partial desensitization of mast cells in the gastrointestinal tract resulting in milder symptoms [2,18].

Allergic eosinophilic esophagitis is usually T-cell-mediated rather than IgE-mediated, and caused by allergens in the diet and, less commonly, in the air [32,34,36-38]. Allergic eosinophilic esophagitis occurs mainly in children and young adults [39-41]. This condition is being more frequently diagnosed over the past decade. It appears that the increase in prevalence is real and is not due to an increased awareness of this condition among physicians. The condition is more common in males and those with a family or personal history of atopy or proven food allergy [33,42,43]. Allergic eosinophilic esophagitis may present with irritability, sleep disturbance, food refusal, vomiting/ regurgitation, dysphagia, abdominal pain, substernal chest pain, occult blood loss, anemia, and failure to thrive [21,25,31,44,45]. Dauer et al. retrospectively reviewed the records of 71 children with biopsy-proven allergic eosinophilic esophagitis and found that the most common symptom was dysphagia, being present in 36 (51%) patients [31]. Eighteen (50%) of the 36 patients with dysphagia experienced at least one episode of food impaction. Other common symptoms include vomiting (31%) and abdominal pain (24%). The symptoms of recurrent vomiting /regurgitation may mimic those of gastroesophageal reflux but are refractory to anti-reflux treatment [25,45,46]. These symptoms do respond to dietary avoidance of food allergens [47]. Typically, affected patients have a negative pH probe study [48].

Allergic Eosinophilic Gastroenteropathy

Allergic eosinophilic gastroenteritis is characterized by infiltration of the gastrointestinal tract with eosinophils, peripheral eosinophilia, and absence of vasculitis [13,49,50]. The eosinophilic infiltrates may be quite patchy and may involve the mucosa, muscular layer, or serosal layer of the stomach or small intestine [24]. In addition to an IgE-mediated mechanism, cell-mediated immunity may also be operative [51]. T-cells specifically sensitized to antigens may release lymphokines capable of attracting eosinophils to the gastrointestinal tissue.

Although allergic eosinophilic gastroenteropathy may affect all ages, the disease is more common in individuals in the third through fifth decades [49]. Patients with mucosal involvement usually have postprandial nausea, vomiting, abdominal pain, watery diarrhea with or without blood, iron deficiency anemia, occasionally steatorrhea, and weight loss in adults or failure to thrive in children [44,52,53]. Patients with muscular involvement may have symptoms and signs of gastric outlet or intestinal obstruction, depending on the site of bowel involvement [24]. The serosal form is characterized by ascites, abdominal pain, and abdominal distension and is extremely rare in children [24,54].

Food Protein-Induced Enteropathy

Food protein-induced enteropathy is characterized by protracted diarrhea and vomiting with onset usually in infancy. This may result in malabsorption, protein-losing enteropathy, and failure to thrive [21,30]. The disorder is caused by a T-cell-mediated response most commonly to cow's milk protein [24]. Intestinal biopsy typically reveals a patchy villous atrophy, increased crypt length, and prominent mononuclear round cell infiltrates [8].

Food Protein-Induced Enterocolitis Syndrome

Food protein-induced enterocolitis syndrome is a cell-mediated hypersensitivity disease which occurs mainly in infants under 3 months of age [4]. The condition usually resolves by

2 years of age but may, rarely, persist into late childhood [25]. Although cow's milk and soy protein are most often responsible [21,33,54], other food antigens have occasionally been implicated, including food protein antigens passed in the maternal breast milk [5,28,55,56]. Classic symptoms are protracted vomiting and diarrhea with progression to dehydration and shock in 20% of patients [30,44,57]. Some infants may have irritability, lethargy, anemia, transient methemoglobinemia, abdominal distension, protein-losing enteropathy, and failure to thrive [21,24,57-59]. Stools generally contain occult blood, polymorphonuclear neutrophils, eosinophils, and Charcot-Leyden crystals [2]. Presumably, stimulation of T-cells by food allergens with secretion of tumor necrosis factor-α may play a role in the pathophysiology of this disorder [30,57,60]. A relative lack of expression of transforming growth factor-β may also have a role to play [57,60]. Skin tests for the offending antigen are usually negative [25]. Radioallergosorbent (RAST) assay, which detects specific IgE antibody, may also be negative in these patients. Jejunal biopsy specimens usually reveal villus atrophy and increased numbers of lymphocytes, eosinophils, and mast cells [61]. Colonoscopy and biopsy show inflammatory colitis and eosinophilic infiltration [55]. Symptoms usually resolve in 72 hours after the offending food substance has been removed from the diet.

Food Protein-Induced Proctocolitis

As with food protein-induced enterocolitis syndrome, food protein-induced proctocolitis is a T-cell mediated disorder. The disorder usually occurs in the first few months of life and is most often secondary to milk protein or soy protein hypersensitivity [44,54,61]. Approximately 60% of cases occur in breastfed infants [5]. Unlike food protein-induced enterocolitis syndrome, infants with food protein-induced proctocolitis generally appear healthy and have normal weight gain. These infants usually have occult or gross blood and sometimes mucus in their stools but they do not have diarrhea [30,62,63]. In general, hematochezia resolves within 72 hours of appropriate food-allergen avoidance [30]. In breastfed infants, elimination of food-allergen from the maternal diet may result in resolution of hematochezia. Colonic biopsy samples reveal mucosal edema, erythema, friability, ulceration, nodular lymphoid hyperplasia and eosinophilic infiltration [44,53,64]. Skin tests for the offending agents as well as RAST assays are usually negative in these patients.

Gluten-Sensitive Enteropathy

Gluten-sensitive enteropathy (celiac disease) is a disorder in which small-bowel mucosal damage is the result of a permanent sensitivity to gliadin, the alcohol-soluble portion of gluten, present in wheat, barley, and rye. Patients with gluten-sensitive enteropathy typically present with diarrhea/steatorrhea, abdominal distension, muscle wasting, and failure to thrive [1]. Other clinical manifestations include irritability, anorexia, vomiting, abdominal pain, oral ulcers, digital clubbing, delayed puberty, and infertility [65,66]. Occasionally, patients with gluten-sensitive enteropathy may be asymptomatic [67]. The presence of anti-gliadin, anti-endomysial, and anti-tissue transglutaminase antibodies of the IgA isotype and anti-gliadin antibodies of the IgG isotype support the diagnosis [21]. However, anti-gliadin antibody of IgG subtype has been known to be positive in conditions such as cow's milk protein allergy, inflammatory bowel disease, and cystic fibrosis and therefore has poor specificity. In addition, patients with IgA deficiency may not have antibodies of IgA subtype in spite of suffering from gluten-sensitive enteropathy. It is recommended that the diagnosis of gluten-sensitive enteropathy be confirmed by intestinal biopsy before instituting dietary changes. Characteristically, biopsy of the jejunum shows villus atrophy, marked increase in crypt-villous ratio, and extensive cellular infiltrates [44]. Both cellular and complement-mediated cytotoxicity and lymphokine-induced damage have been implicated in the pathogenesis of the condition [47]. There is a genetic predisposition to the disease. There is a predominance of certain HLA types (B8, DQ2, DW3) in patients with gluten-sensitive enteropathy [1,2,68]. Environmental factors may influence expression of the genetic predisposition.

Infantile Colic

There is increasing evidence that cow's milk proteins may play an important role in the pathogenesis of infantile colic [69-75]. Approximately 25% of infants with moderate or severe colic have allergy to cow's milk protein [76,77]. Lothe and Lindberg showed that colic disappeared in 24 of 27 infants when they were given a cow's milk-free diet [73]. These infants were entered into a double-blind placebo-controlled crossover trial of whey protein. Eighteen infants receiving the whey protein capsules reacted with colic, two infants received placebo reacted with colic, and four infants did not react at all.

Iacono et al. put 70 cow's milk formula-fed infants with severe colic on soy-milk formula [71]. In 50 infants, there was a remission of symptoms when cow's milk protein was eliminated from their diet. Two successive challenges caused the return of symptoms in all these 50 infants. Follow-ups, after an average period of 18 months, showed that in 22 of 50 (44%) of the infants who had cow's milk protein-related colic and 1 of 20 (5%) of those with non-cow's milk protein-related colic developed an overt form of alimentary intolerance. Lucassen et al randomly selected 43 healthy infants with colic to receive whey hydrolysate formula or standard formula [74]. They found a decrease in the duration of crying in those infants fed with whey hydrolysate formula.

Jakobsson et al. studied the effectiveness of 2 formulae with extensively hydrolysed casein in 22 infants with severe colic [78]. One infant was considered as treatment failure and six infants as protocol failures. The remaining 15 infants showed a significant decrease in the lengths of time they cried as well as a decrease in the intensity of their crying on both formulae. When the infants were challenged in a double-blind design, 11 infants reacted with an increase in crying time to cow's milk protein or bovine whey protein.

Hill et al. studied the effect of diet change in 38 bottle-fed and 77 breast-fed colicky infants in a double-blind, randomized, placebo-controlled trial [70]. Bottle-fed infants were assigned to either casein hydrolysate or cow's milk formula. All mothers of breast-fed infants were started on an artificial color-free, preservative-free, additive-free diet and were randomized to receive either an active low allergen (milk free) diet or a control diet. Hill et al. showed that infants on the active diet had their distress reduced by 39% compared with 16% for those on the control diet.

Jakobsson and Lindberg put 66 mothers of 66 breast-fed infants with infantile colic on a cow's milk-free diet [79]. The colic disappeared in 35 infants; it reappeared after reintroduction of cow's milk into the mother's diet in 23 of the 35 infants. A double-blind crossover trial with cow's milk whey protein was performed on 16 of these 23 mothers and infants. Six infants had to be taken out of the study for various reasons. Of the remaining 10 infants, nine displayed signs of colic after their mothers had taken the whey-filled capsules.

Maternal ingestion of eggs, chocolate, citrus fruits, nuts, as well as certain seafood whilst breastfeeding may result in infantile colic [72,80-82]. Hill et al. randomized mothers of 107 term breastfed infants younger than 6 weeks of age with colic to follow a low-allergen diet with elimination of dairy products, soy, wheat, eggs, peanuts, tree nuts, and fish (n=53) and a control group (n=54) whose diet contained the known allergen [83]. Forty seven mothers in the treatment group and 43 mothers in the control group completed the study. Infants were identified as responders if there was at least 25% reduction in duration of crying/fussing on days 8 and 9. The authors showed that 74% of infants in the treatment group versus 37% of infants in the control group were responders (p=<0.001).

Irritable Bowel Syndrome

The pathogenesis of irritable colon syndrome is likely heterogeneous. Food allergy has been implicated in the pathogenesis of a subset of patients with irritable bowel syndrome [84,85]. The association of irritable bowel syndrome with specific IL-10 genotypes supports involvement of the immune system in its pathogenesis [58,86,87]. It has been shown that patients with irritable bowel syndrome have a greater area of intestinal mucosa occupied by mast cells than do healthy control individuals [88]. A study of dietary eliminations in patients with irritable bowel syndrome found a significant reduction in symptom score in those patients whose exclusions were guided by raised IgG antibodies to dietary antigens than did patients on a sham diet based on irrelevant antigens [89]. Reintroduction of eliminated foods resulted in aggravation of symptoms in patients whose dietary exclusions were guided by raised IgG antibodies to dietary antigens. Irritable bowel syndrome might result from an interplay between immunological dysfunction, impaired gut barrier functions, susceptible genes and other environmental factors [90]. It has been hypothesized that food antigens induce mast cells to secrete mediators that regulate gastrointestinal motility and pain perception through gastrointestinal neural system [84].

Recurrent Abdominal Pain

Recurrent abdominal pain is usually defined as three or more bouts of abdominal pain, severe enough to interfere with a child's normal activities, occurring over a period of not less than three months during the year preceding the examination [91,92]. Onset often occurs between five and 10 years of age. Typically, the pain is vague, poorly localized or periumbilical and may be crampy or sharp. Episodes of pain tend to cluster, alternating with pain-free periods of variable length [92]. Most episodes last for less than an hour. On cessation of the pain, the child is up and about as if nothing had happened. In the majority of cases, the cause is functional [91,92]. Organic causes account for 5 to 10% of cases [91,92].

Kokkonen et al. studied 84 children with recurrent abdominal pain [93]. Food allergy was diagnosed in 28 (33%) children based on an open elimination challenge test. The study was criticized because a formal diagnosis would require a double-blind placebo-controlled food challenge [94]. Further studies using double-blind placebo-controlled food challenges are necessary before food allergy can be established as a cause of recurrent abdominal pain in children.

Constipation

The vast majority of constipation in children is functional [95,96]. Constipation resulting from IgE sensitization to cow's milk has been described [97]. Loening-Baucke reviewed the records of 4,157 children between 0 to 24 months of age seen in general pediatric clinics for health maintenance and acute care visits [96]. Of the 185 children with constipation, food allergy was responsible for constipation in only 2 (1%) of these children [96]. In contrast, Iacono et al studied 27 consecutive infants with chronic "idiopathic" constipation and noted improvement or resolution of symptoms in 21(78%) of these infants after a 4-week period of a cow's milk-free diet [97]. These infants had a relapse of symptoms on cow's milk challenge. Iacono et al. subsequently performed a randomized cross-over trial of a cow's milk-based diet versus a soy milk-based diet in 65 children with chronic constipation [98]. Forty-four (68%) of the 65 children had increased bowel movements and improvement of fecal score while receiving the soy milk. None of the children who received cow's milk had a response. In all 44 children with a response, the response was confirmed with a double-blind challenge with cow's milk. Daher et al studied 25 children with chronic constipation [99]. In seven (28%) patients, the constipation disappeared while they were following a diet free of cow's milk protein and reappeared within 48 to 72 hours after challenge with cow's milk. In two patients, a rectal biopsy revealed allergic colitis with eosinophilic infiltration and they therefore did not undergo the challenge. High serum levels of total IgE were observed in five (71%) of the seven patients who showed a clinical improvement. Two (29%) patients had a positive skin test and two (29%) had detectable levels of specific IgE. Carroccio et al treated 52 children with chronic constipation unresponsive to common treatment by exclusion of milk alone, or by an extensive oligoallergenic diet if unresponsive [100]. Twenty four patients were found to be suffering from cow's milk intolerance and six from multiple food intolerance. These patients had a normal stool frequency on elimination diet with recurrence of constipation on food challenge. These patients showed a significantly higher frequency of mucosal erosions, number of intraepithelial lymphocytes and eosinophils, and number of eosinophils in the lamina propria. The remaining 22 patients did not respond to the elimination diet. Murch identified 30 children with constipation who responded to the exclusion diet with resolution of symptom; six of these children were allergic to multiple antigens [58]. Rectal biopsy of the affected patients showed eosinophilic proctitis [58]. Carroccio et al. performed a Medline search for articles published between 1970 and June 2006, using the key words "chronic constipation or constipation" and "food intolerance or allergy". The authors identified 33 papers but only 19 of them were related to the topic. Analysis of these papers showed a relationship between constipation and food allergy in a subgroup of pediatric patients with "idiopathic" constipation unresponsive to laxative treatment. Additional studies are necessary to substantiate the specific associations and to clarify the pathogenic mechanisms involved.

Cutaneous Manifestations

Urticaria/Angioedema

Acute urticaria and, to a lesser extent, angioedema are among the most common manifestations of food allergic reactions in children [29,44,52]. They tend to occur more commonly in younger patients and in atopic patients [101]. Symptoms result from activation of IgE-bearing cells by circulating food allergens absorbed through the gastrointestinal tract. The foods most commonly incriminated include eggs, milk, peanuts, and nuts [102]. In several studies, urticaria/angioedema occurred in 10 to 15% of infants with challenge-proven milk allergy [103,104]. In a study of 554 adults with urticaria, food allergy was demonstrated as the cause in only 1.4% of cases [105].

Even contact with foods may also cause acute urticaria [104,106,107]. Allergic contact urticaria can be seen in children who are sensitized to environmental allergens such as food or latex allergy [108]. There is a potential for cross-reactivity with various foods in individuals with latex allergy [108]. Food allergy is rarely the cause of chronic urticaria, unless the offending food is eaten almost every day [28,44,52,101].

Atopic Dermatitis

Atopic dermatitis is a multifactorial disease, and food allergy plays an immunopathogenic role in 30 to 50% of young children who had moderate to severe atopic dermatitis [5,109-111]. Burks et al. evaluated 46 patients who had atopic dermatitis from food hypersensitivity substantiated with double-blind placebo-controlled food challenges [112]. Sixty five food challenges were performed; 27 (42%) were interpreted as positive in 15 (33%) patients. Sampson et al. studied 350 patients with severe atopic dermatitis for possible food hypersensitivity [28,113,114]. Food allergy was diagnosed by double-blind placebo-controlled food challenges. Cutaneous reactions developed in 75% of the positive challenges within minutes to two hours, but only 30% of the positive responses were isolated cutaneous symptoms alone [28]. Most of the skin manifestations consisted of a markedly pruritic, erythematous rash that developed in sites with a predilection for atopic dermatitis.

In a well-designed prospective study of 113 patients with atopic dermatitis, marked improvement was noted in those who were maintained on an allergen elimination diet when compared with a similar group of patients who did not have food allergy or who did not adhere to the elimination diet [113]. Breuer et al. performed 106 double-blind placebo-controlled food challenges to cow's milk, egg, wheat, and soy on 64 children who had atopic dermatitis [115]. Twenty-eight (57%) of the 49 positive reactions resulted in late eczematous reactions either as isolated events or in combination with immediate reactions. Hill et al. evaluated 487 infants who had skin prick tests to cow's milk, egg, and peanut, and who had a family history of atopic dermatitis, asthma, or immediate food allergy in a parent or sibling

[116]. One hundred and forty-one (28.9%) of these infants had atopic dermatitis by the age of 12 months. These authors found that as the severity of atopic dermatitis increased, so did the prevalence of IgE-mediated food allergy and also the frequency of adverse food allergy reactions. The relative risk of an infant who had atopic dermatitis to develop an IgE-mediated food allergy was 5.9 for the most severely affected group. The most frequently implicated foods include eggs, cow's milk, tree nut, peanut, soy, wheat, seafood, citrus fruits, and chocolate [117]. Approximately 30 to 40% of children lose their food hypersensitivity after one to two years of allergen avoidance and 80 to 85% outgrow their food allergies by 10 years of age [81,117]. Hypersensitivity to peanut, tree nut, and shellfish is more persistent [117].

The pathogenesis of atopic dermatitis likely involves both immediate and late-phase effects of IgE-mediated food hypersensitivity reactions as well as cell-mediated reactions [28,29,52,106]. The immediate or early phase of the reaction results from IgE-mediated cutaneous mast cell activation [29,114]. The late phase is characterized by a mixed cellular infiltrate (eosinophils, neutrophils, lymphocytes, and basophils) at six to eight hours and thereafter by a mononuclear round cell infiltrate indistinguishable from that seen in eczematous skin [113,118,119]. The pattern of cytokine expression is predominantly that of the Th2 type, namely, interleukin-4, -5 and -13 [101,102]. A single ingestion of food allergen may not provoke an eczematous lesion, but chronic ingestion of a food allergen can result in the classic changes of atopic dermatitis [109]. Children who have atopic dermatitis and documented food allergy might develop typical eczematous lesions while the disease is active, but might develop urticaria with ingestion of a food allergen when the atopic dermatitis is in remission [111]. On the other hand, atopic dermatitis is also seen in patients with X-linked agammaglobulinemia, suggesting that, at least in some cases, atopic dermatitis is not IgE-mediated [120]. Testing for cow's milk allergy, Hill et al. identified a delayed eczematous reaction in 17 of 135 children with atopic dermatitis [121]. Ten of the 17 children had negative prick test to cow's milk allergen, suggesting a non-IgE mediated pathogenesis. Lio suggests that there are at least two types of "food allergy" in patients with atopic dermatitis: the IgE-mediated immediate reactions and the cell-mediated eczematous reactions [120].

Contact Dermatitis

Contact dermatitis may be related to an immune-mediated reaction to food or a direct toxic effect of the food coming into contact with the skin [102,122]. Food-induced contact dermatitis is often seen among food handlers, especially those who handle raw shellfish and eggs [102,123,124]. Allergic contact cheilitis has been reported from the chewing of garlic [125] and from the contact of geraniol, a food additive contained in certain foods [126].

Dermatitis Herpetiformis

Gluten-sensitive enteropathy is found in 75 to 95% of patients with dermatitis herpetiformis [106,127]. Dermatitis herpetiformis is a cutaneous cell-mediated response to gliadin which is present in wheat, rye, and barley [44,102]. The disorder is characterized by a chronic, intensely pruritic, papulovesicular rash, symmetrically distributed over the extensor surfaces of the extremities and buttocks [8,102]. Approximately 80 to 90% of patients have the HLA-B8 haplotype, and more than 90% have the HLA-Dw3 haplotype [29,128]. Skin biopsy generally reveals granular and linear deposits of IgA, C3, as well as infiltrates with polymorphonuclear leukocytes. Immunoglobulin A deposits may activate complement through the alternative pathway and cause inflammation [29]. An IgE-mediated hypersensitivity reaction does not contribute to the pathogenesis. IgA antibodies to smooth muscle endomysium have been reported in patients with dermatitis herpetiformis-associated gluten-sensitive enteropathy [106]. Also, antibodies to tissue transglutaminase and epidermal transglutaminase are present in patients with dermatitis herpetiformis, suggesting that epidermal transglutaminase is the autoantigen in dermatitis herpetiformis [102].

Respiratory Manifestations

Rhinitis/ Rhinoconjunctivitis

Food-induced allergic rhinitis/rhinoconjunctivitis is more frequently observed in children than in adults [110,129]. Rhinitis/rhinoconjunctivitis following inhalation of food dusts or vapor is not uncommon in patients with food allergy [10]. Rhinitis/rhinoconjunctivitis typically occurs in association with other clinical manifestations such as cutaneous and/or gastrointestinal symptoms during acute allergic reactions to foods [19,130]. Rhinitis/ rhinoconjunctivitis as the sole manifestation of food allergy is quite uncommon [28,54,124,129]. Allergic rhinitis may be manifested as nasal congestion, sneezing, and rhinorrhea [117]. Allergic conjunctivitis is characterized by periocular erythema, ocular injection, pruritus, and tearing. It seems likely that ingested allergens can activate nasal mast cells in addition to mast cells elsewhere in the body [29].

Chronic Sinusitis

Allergy to food allergens has been suggested to be a rare cause of chronic sinusitis [131]. Food allergy should be suspected in refractory cases of chronic sinusitis in which no apparent cause can be found, especially in atopic individuals with perennial symptoms [108,131].

Asthma

Asthma from food allergy is a mixed IgE- and cell-mediated response due to the involvement of IgE antibodies, mast cells, eosinophils, and T-lymphocytes [21]. The condition occurs more often in children than in adults [132,133]. In several studies, 2 to 6% of children with asthma were found to have wheezing provoked by blinded food challenges [129,133-135]. Asthmatic children have a 14-fold higher risk of developing a severe allergic reaction to food compared with children without asthma [133,136].

In some children, food allergy may increase airway reactivity so that other triggers or environmental factors can more readily precipitate an asthmatic attack [137,138]. Even vapors containing proteins emitted from cooking food can induce asthmatic attacks [124,132,139,140]. Inhalational exposures to foods, particular in the workplace, account for approximately 1% of asthmatic attacks in the adult population [124,138]. Baker's asthma caused by inhalation of flour or mold-derived enzymes used as flour additives is a good example [44]. Affected individuals have asthmatic attacks in association with exposure to aerosolized wheat proteins and have positive skin prick tests or serum specific IgE to wheat proteins [135,138]. Likewise, peanut dust in airplanes can provoke allergic reactions in susceptible individuals [141]. Allergenic proteins may also reach the respiratory tract via the circulation or they may act via inflammatory mediators released from the skin or gastrointestinal tract [138].

Food-induced asthmatic reactions, when they occur in conjunction with other organ systems, generally indicate a more severe disease manifestation [132,142,143]. In a survey of six fatal and seven near fatal anaphylactic reactions after food ingestion, all patients had asthma and respiratory symptoms as part of their clinical manifestations [109]. Food allergy in early infancy increases the risk for developing asthma in later life [133,144,145]. This applies also to children who have outgrown their food allergies [142].

Heiner Syndrome

Primary pulmonary hemosiderosis with hypersensitivity to cow's milk (Heiner syndrome) is a rare condition that usually occurs in young children. The syndrome is characterized by chronic cough, wheezing, hemoptysis, pulmonary infiltrates, hemosiderosis, gastrointestinal blood loss, iron deficiency anemia, and failure to thrive [129,146]. Presumably, the syndrome results from aspiration of milk into the lungs with subsequent development of IgG cow's milk antibodies and an immune complex Arthus-type reaction in the alveoli [68,147]. If the vasculitis is severe, alveolar bleeding and pulmonary hemosiderosis may result [147]. Affected children usually have high titers of precipitins to multiple constituents of cow's milk and positive results on intradermal skin tests to various cows' milk proteins [146]. Dietary elimination of cow's milk results in symptomatic improvement, and reintroduction of cow's milk results in recurrence of symptoms. It has been postulated that antigen-antibody complexes and cell-mediated hypersensitivity play a pathogenic role in this syndrome based on the presence of elevated serum levels of milk-specific IgG antibodies and the in vitro proliferative response of the patient's lymphocytes to milk antigen [29].

Serous Otitis Media

Serous otitis media, defined as non-purulent collection of fluid in the middle ear, is a multifactorial disease [148]. Food allergy may play a role in the pathogenesis in a subgroup of children with serous otitis media [149-151]. It has been hypothesized that allergic inflammation in the nasal mucosa may cause Eustachian tube dysfunction which may result in serous otitis media [130,150]. IgG complexes with cow's milk protein might also contribute to the middle ear inflammation [152]. In the study by Aydoğan et al., food allergy was detected in 25 (44.6%) of 56 patients with serous otitis media [148]. In patients with food allergy, serous otitis media was detected in 7 (25%) of 28 patients. In the control group of 28 patients, food allergy was diagnosed in 5 (18%) patients and serous otitis media in one (3%) patient. The incidence of food allergy in serous otitis media group was statistically significant when compared to the normal group ($p<0.05$). The risk of otitis media in children having food allergy was 3.7 times higher than the control ($p<0.05$). Bernstein et al evaluated 100 patients aged 2 to18 years with recurrent serous otitis media for IgE-mediated hypersensitivity [149]. Thirty five of these patients had allergic rhinitis. Total IgE was increased in the middle ear fluid in 16 of the 35 patients with allergic rhinitis and in only 2 of the 65 patients in the non-allergic group. In 8 of these 16 patients (23% of the allergic group), levels of IgE per milligram of protein were higher in the middle ear effusion than in the corresponding serum, thus suggesting local production of IgE by the middle ear mucosa in these patients. In a subset of infants with recurrent serous otitis media, IgG complexes with food antigen may contribute to middle ear inflammation and serous otitis media [130,152]. Nsouli et al. evaluated 104 unselected children with recurrent serous otitis media for food allergy [151]. There was a significant statistical association, by chi-square analysis, between food allergy and recurrent serous otitis media in 81 (78%) of the 104 patients. An elimination diet led to a significant amelioration of serous otitis media in 70 (86%) of 81 patients. An open challenge diet with the suspected offending food(s) provoked a recurrence of serous otitis media in 66 (94%) of 70 patients. Further studies with double-blind placebo-controlled food challenge are necessary to confirm these findings.

Ménière's Disease

Ménière's disease is characterized by recurrent vertigo, fluctuating hearing loss, aural fullness or pressure, and tinnitus [153,154]. Derebery et al. did a mail-survey on 1490 patients with Ménière's disease [153]. Of 734 respondents with Ménière's disease, 296 (40.3%) had or suspected food allergies and 272 (37%) had confirmatory skin or in vitro tests for allergy. These prevalence rates were significantly higher than those found in the control group of patients ($n=172$) with otologic problems other than Ménière's disease, of which 43 (25%) had or suspected food allergies and 38 (22.2%) had confirmatory skin or in vitro tests for allergy (differences all significant at $p\leq0.005$). Keleş et al. studied 46 patients aged between 26 and 68 years and 46 age-matched controls. The authors found that total serum IgE levels were above the normal levels in 19 (41.3%) of the patients with Ménière's disease and 9 (19.5%) of the controls [154]. A history of allergy was found in 31 (67.3%) of the patients with Ménière's disease and 16 (34.7%) of the controls. When the specific IgE levels were measured (all seasons, tree, fungus, fruit, egg white, cow's milk, wheat flour, corn flour, beef, and rice), the number of patients having all the panels negative was eight (17.9%) in patients with Ménière's disease and 31 (67.3%) in the controls. Cow's milk allergy was the most common identifiable cause of Ménière's disease. It has been hypothesized that immune complexes circulating in the blood might hinder the filtering ability of the endolymphatic sac [154].

Generalized Manifestations

Systemic Anaphylaxis

Systemic anaphylaxis is an uncommon but potentially fatal manifestation of food allergy [109]. Systemic anaphylaxis usually occurs within minutes, but occasionally hours after the ingestion of an offending food [155]. Peanuts, nuts, eggs, and seafood are responsible for the majority of these reactions [65,156]. Systemic anaphylaxis results when antigen binds to allergen-specific IgE on mast cells and basophils with sudden release of potent biologically active inflammatory mediators affecting multiple target organs [45]. The mediators include histamine, heparin, and tryptase [7]. Early symptoms may include pruritus, "metallic" taste in the mouth, sensation of tightness in the throat, flushing, urticaria, dizziness, nausea, vomiting, abdominal pain, angioedema, and wheezing [157]. This may rapidly progress to laryngeal edema, dyspnea, stridor, diaphoresis, cyanosis, chest pain, hypotension, cardiac dysrhythmias, and shock [52,158]. The degree of anaphylactic reactions varies and may be manifested in a partial form [12]. In general, the more rapidly anaphylaxis occurs after exposure to an offending agent, the more likely the anaphylactic reaction is to be severe and potentially life-threatening [156]. Anaphylactic reactions to foods can be biphasic with an early and late phase separated by one to eight hours or there may be multiple recurrences separated by asymptomatic periods lasting for hours [10,68]. Some very severe anaphylactic reactions are protracted and last continuously for many hours without remission [68]. Risk factors for severe anaphylactic reactions include history of a previous anaphylactic reaction, history of poorly controlled asthma, allergy to peanuts, nuts and shellfish, and use of ß-blockers or acetycholinesterase inhibitors [1,159]. Low levels of serum platelet-activating factor acetylhydrolase may be a marker for more severe food-induced anaphylaxis [159].

Food-Dependent Exercise-Induced Anaphylaxis

Anaphylaxis has been reported after the ingestion of foods in association with exercise [160-162]. Food-dependent exercise-induced anaphylaxis represents 7 to 9% of anaphylactic reactions [163,164]. The condition is twice as common in females and 60% of cases occur in individuals under the age of 30 years [157,159,163]. There is often a history of atopy [163]. One subset of patients may develop anaphylaxis in temporal proximity to ingestion of any type of food [44,157]. The other subset may develop anaphylaxis with exercise in conjunction with ingestion of a specific food [157]. The latter subset is more common than the former subset [157]. Foods associated with food-specific exercise-induced anaphylaxis include crustaceans, celery, grapes, tomato, wheat, buckwheat, chicken, and dairy products [44,165]. Rarely, two foods have to be eaten together to provoke an anaphylactic attack [167]. When each food is taken separately, food-dependent exercise-induced anaphylaxis does not occur [167]. Typical symptoms include urticaria, angioedema, dyspnea, and abdominal pain [168]. These may progress to hypotension and shock. Loss of consciousness is seen in approximately 30% of cases [168].

Although various exercises may lead to anaphylaxis in susceptible individuals, jogging is the exercise most frequently reported, followed by aerobics and walking [44,163,169].

Anaphylaxis usually occurs when exercise takes place within two to four hours of food ingestion [157]. Unlike exercise-induced anaphylaxis, anaphylactic symptoms develop only in the presence of both food ingestion and exercise [170]. Food-dependent exercise-induced anaphylaxis often presents with pruritus about the scalp before the symptoms become generalized [157].

The exact mechanism of food-dependent exercise-induced anaphylaxis is not known. There is evidence of IgE-mediated sensitization to the food allergen [163]. Skin testing may show an immediate flare-and-wheal reaction to the implicated food [162]. Blood flow differences to the gut, increased food allergen absorption, increased spontaneous leukocyte histamine release, lowered mast cell releasability threshold, and enhanced mast cell responsiveness to physical stimuli may have a role in the pathogenesis of this condition [44,162,169,171].

Hyperactivity

A few anecdotal reports have suggested a role of food allergy in some children with hyperactivity [172,173]. To date, controlled studies have not substantiated the efficacy of an elimination diet in the treatment of hyperactivity [174]. The placebo effect may account for the favorable results in some of the uncontrolled studies [174-176].

Neurologic Manifestations

Migraine

Several double-blind studies have shown that some patients with migraine had adverse reactions to certain foods, as shown by dietary exclusion and subsequent challenge [177-179]. There was no clearly proven immunologic effect of food components in many of these patients [12]. Migraine can be triggered by foods rich in tyramine, tryptamine, serotonin, and histamine, and this is based mainly on pharmacologic rather than immunologic effects [180,181]. According to the revised nomenclature for allergy published by the European Academy of Allergy and Clinical Immunology, these kinds of adverse reactions are classified as toxic reactions rather than food hypersensitivity [182]. The present consensus is that food allergy does not play a significant role in migraine [110].

Epilepsy

Several authors have proposed a possible role of food allergy in a subset of children with epilepsy [183-185]. Frediani et al. have noted a significantly higher proportion of allergic disorders in 72 epileptic patients versus 202 aged-matched controls [184]. Using skin prick tests, the percentage of epileptic children who tested positive for cow's milk protein (24/72) and especially for lactalbumin (16/72) and β-lactoglobulin (10/72) was significantly higher than the percentage of controls that tested positive for the same allergens: 7/202 for cow's milk protein, 4/202 for lactalbumin, and 2/202 for β-lactoglobulin. In patients with allergic disorders, there is an increase in the proportion of electroencephalographic anomalies, often in the form of occipital dysrhythmias [184]. Crayton et al. reported on a patient whose epileptic fits were triggered and increased in frequency by double-blind food challenges [183]. Frediani et al. reported on a nine-year-old boy whose epileptic symptoms disappeared as a result of an allergen-free diet with no anticonvulsant therapy [185].

Endocrine Manifestation

Diabetes Mellitus

Cow's milk has been implicated as a possible trigger of the autoimmune response that destroys pancreatic ß-cells in genetically susceptible individuals, thereby leading to diabetes mellitus [186-188]. Karjalainen et al. measured IgG antibovine serum albumin antibodies in the serum of 142 children with newly diagnosed insulin-dependent diabetes mellitus [188]. These children had elevated serum concentrations of IgG antibovine serum albumin antibodies that declined after diagnosis and reached normal levels in most patients within one to two years. These authors speculate that patients with insulin-dependent diabetes mellitus have immunity to cow's milk albumin, with antibodies to an albumin peptide that are capable of reacting with a β-cell-specific surface protein. Such antibodies could participate in the development of islet cell dysfunction. Cavallo et al. measured the in vitro peripheral T-lymphocyte response to β-casein in 47 patients with recent-onset insulin-dependent diabetes mellitus [186]. Twenty four of 47 (51.1%) patients with insulin-dependent diabetes mellitus versus 0 of 10 patients with autoimmune thyroid disease and 1 of 36 (2.8%) healthy people had a positive response to ß-casein. A positive response was defined as a stimulation index above the mean value plus 2 SD of healthy people. These authors suggest that exposure to cow's milk triggers a cellular and humoral anti- ß-casein immune response that may cross-react with a β-cell antigen and lead to destruction of the β-cell. A subsequent study, however, showed a lack of association between early exposure to cow's milk and β-cell autoimmunity [189]. Norris et al. screened 253 children aged nine months to seven years with first-degree relatives who had insulin-dependent diabetes mellitus for β-cell autoimmunity [189]. Eighteen cases of β-cell autoimmunity were detected at baseline. These children were compared with 153 unrelated autoantibody-negative children selected from the cohort as controls. There were no differences in the proportion of cases and controls that were exposed to cow's milk or foods containing cow's milk by 3 months or 6 months of age. Further studies are necessary to clarity this important issue.

Renal Manifestations

Nephrotic Syndrome

Food allergy may play a role in the pathogenesis of nephrotic syndrome in a selected group of children [190-193]. Sandberg et al. reported 6 children with steroid-responsive nephrotic syndrome [190]. In the relapse period, a milk-free diet led to remission without steroid therapy. In the remission period, challenge with cow's milk resulted in a relapse in 4 patients. Sieniawska et al. evaluated the role of cow's milk in 17 children with steroid-resistant nephrotic syndrome [193]. Cows' milk was excluded from the diet for at least 14 days without changing the previously ineffective prednisone dosage. Six patients went into remission three to eight days after the elimination of cow's milk. After a period of two to three weeks of remission, cow's milk challenge was positive in three of the six patients. The group of responders to a milk-free diet was characterized by young age, feeding with cow's milk or unmodified powdered milk formulas in the neonatal period, and coexistence of allergic symptoms. The authors suggest that cellular mechanisms may play a role in cow's milk-induced steroid-resistant nephrotic syndrome as evidenced by the late-onset reaction to cow's milk challenge, positive leukocyte migration inhibition tests, absence of specific IgE antibodies, and negative skin test results. These studies, however, contain some design flaws and better objective studies are needed to prove the association between milk ingestion and nephrotic syndrome.

Nocturnal Enuresis

Several authors have noted that food allergy might play a role in nocturnal enuresis [194-196]. Mungan et al. studied an allergy panel that included total IgE, 10 examples of inhalant-specific IgE, 10 examples of food-specific IgE, eosinophilic cationic protein and Phadiotop on 37 children with nocturnal enuresis and 18 children without nocturnal enuresis as a control [94]. The authors did not find statistically significant differences between the two groups in terms of levels of total IgE, the 10 examples of inhalant-specific IgE and Phadiotop. However, two (soybean and hazelnut) of the 10 food-specific IgE and eosinophilic cationic protein level did differ significantly between the two groups. Further studies are necessary before a causal relationship can be established.

Hematologic Manifestations

Anemia

Iron deficiency anemia may develop in children with cow's milk allergy secondary to gastrointestinal blood loss. This may be caused by milk-induced enterocolitis syndrome [54], milk-induced colitis, allergic eosinophilic gastroenteropathy [52], and Heiner syndrome [146].

Thrombocytopenia

A few anecdotal reports suggest that thrombocytopenia may be caused by food allergy [197-199]. Whitefield and Barr reported a girl with the syndrome of thrombocytopenia and absent radius who showed marked gastrointestinal disturbance with clinical evidence of cow's milk allergy and in whom there appeared to be a direct correlation between cow's milk exposure, gastrointestinal upset, and thrombocytopenia [197]. Jones reported a newborn male infant with idiopathic thrombocytopenic purpura in whom withdrawal of a milk formula produced an improvement in the platelet count and reintroduction of the milk formula led to hematologic relapse on two occasions [198]. It has been suggested that a type II cytotoxic reaction may account for the thrombocytopenia seen after the ingestion of milk [199].

Cardiovascular Manifestation

Vasculitis

Food-induced vasculitis has been described [200,201]. In the majority of cases, it is mediated by type III immunologic reaction (Arthus type) in which the antigen combines with its specific IgG and complement to form circulating complexes [200]. The circulating complexes deposit in the small blood vessels and initiate vasculitis. Occasionally, food-induced vasculitis may be IgE-mediated. Businco et al. reported two patients with leucocytoclastic vasculitis confirmed by skin biopsy [200]. The first patient had cutaneous vasculitis with large joint involvement, caused by cow's milk and egg as confirmed by blind food challenge. The second patient had cutaneous and mucous membrane vasculitis with large joint involvement caused by chocolate. Lunardi et al. described five patients with allergy and cutaneous vasculitis of 1 to 13 years' duration [201]. Double-blind food challenges identified the offending agent to be a food in two patients, an additive in another two patients, and both food and additive in the fifth patient.

Rheumatic Manifestation

Arthropathy/Arthritis

A few anecdotal reports suggest that arthropathy/arthritis may be the result of food hypersensitivity [202-204]. Parke et al. described a 38-year-old woman who had progressive rheumatoid arthritis for 11 years [205]. Her rheumatoid arthritis improved within three weeks of changing to a milk-free diet and deteriorated within 24 hours with a milk challenge. Golding reported three patients with food-induced synovitis [206]. van de Larr et al. described six patients with rheumatoid arthritis which responded to a diet limited to an elemental formula [207]. Double-blind placebo-controlled trial confirmed a relationship with specific food in four of the six patients. Long-term benefit from avoidance of the specific foods, however, was noted only in two patients. Panush monitored 97 patients with inflammatory arthritis and found that no more than 5% of the patients with rheumatic disease had immunologic sensitivity to foods [204]. Denman et al. have not been able to detect any consistent correlation between controlled dietary challenges and exacerbations of inflammatory arthritis [208]. In those cases in which inflammatory arthritis responds to dietary manipulation, it is possible that dietary restriction non-specifically moderates the inflammatory manifestations of the disease or the placebo effect may be responsible [208].

Karatay et al. have shown that individualized diet challenges consisting of allergenic foods may regulate tumor necrosis factor-α and interleukin-1β levels in selected patients with rheumatoid arthritis [209]. Tumor necrosis factor-α and interleukin-1β are cytokines which promote inflammation and may play an important role in the development of rheumatoid arthritis [209].

Conclusions

The clinical manifestations of food allergy can be protean. There is usually a clear temporal relationship between the allergic reaction and prior exposure to food. At times, recognition of the allergic symptoms can be difficult as they may develop hours or days after food ingestion and the presentation could be atypical or unusual. In addition, the offending agent might not be the main food that was ingested [210]. Awareness of the various clinical manifestations of food allergy is important so that appropriate investigations can be planned and treatment initiated.

References

[1] Burks, W; Ballmer-Weber, BK. Food allergy. *Mol. Nutr. Food Res.* 2006;50:595-603.

[2] Scurlock, AM; Lee, LA; Burks, AW. Food allergy in children. *Immunol. Allergy Clin. North Am.* 2005;25:369-388.

[3] Hyams JS. Food allergy (food hypersensitivity). In: Kleigman, RM; Behram, RE; Jenson, HB; et al. (eds). *Nelson Textbook of Pediatrics*. Philadelphia: Saunders. 2007; pp.1585-1587.

[4] Pirson, F. Food allergy: a challenge for the clinician. *Acta Gastroenterol. Belg.* 2006;69:38-42.

[5] Sampson, HA; Leung, DY. Adverse reactions to foods. In: Kleigman, RM; Behram, RE; Jenson, HB; et al. (eds). *Nelson Textbook of Pediatrics*. Philadelphia: Saunders. 2007; pp. 986-990.

[6] Bangash, SA; Bahna, SL. Pediatric food allergy update. *Curr. Allergy Asthma Rep.* 2005;5:437-444.

[7] Keet, CA; Wood RA. Food allergy and anaphylaxis. *Immunol. Allergy Clin. North Am.* 2007;27:193-212.

[8] Mansueto, P; Montalto, G; Pacor, ML; et al. Food allergy in gastroenterologic diseases: review of literature. *World J. Gastroenterol.* 2006;12:7744-7752.

[9] Sicherer, SH; Sampson, HA. Food allergy. *J. Allergy Clin. Immunol.* 2006;117:S470-S475.

[10] Asero, R; Ballmer-Weber, BK; Beyer, K; et al. IgE-mediated food allergy diagnosis: current status and new perspectives. *Mol. Nutr. Food Res.* 2007;51:135-147.

[11] Minford, AMB; MacDonald, A; Littleword, JM. Food intolerance and food allergy in children: A review of 68 cases. *Arch. Dis. Child.* 1982;57:742-747.

[12] Stern, M. Allergic enteropathy. In: Walker, WA; Durie, PR; Hamilton, JR; et al. (eds). *Pediatric Gastrointestinal Disease*. St Louis: Mosby. 1996; pp. 677-692.

[13] Burks, AW; Sampson, HA. Diagnostic approaches to the patient with suspected food allergy. *J. Pediatr.* 1992;121:S64-S71.

[14] Sugita, K; Kabashima, K; Nakashima, D; et al. Oral allergy syndrome caused by raw fish in a Japanese sushi bar worker. *Contact dermatitis*. 2007;56:369-370.

[15] Nash, S; Burks, AW. Oral allergy syndrome. *Curr. Allergy Asthma Rep.* 2007;7:1-2.

[16] Purohit-Sheth, TS; Carr, WW. Oral allergy syndrome (pollen-food allergy syndrome). *Allergy Asthma Proc.* 2005;26:229-230.

[17] Seidman, E. Food allergic disorders of the gastrointestinal tract. In: Roy, CC; Silverman, A; Allagille, D. (eds). *Pediatric Clinical Gastroenterology*. St Louis: Mosby. 1995; pp. 374-384.

[18] Sampson, HA. Food allergy – accurately identifying clinical reactivity. *Allergy*. 2005;60:19-24.

[19] Aiuti, F ; Paganelli, R. Food allergy and gastrointestinal diseases. *Ann. Allergy*. 1983;51:275-280.

[20] Amlot, PL; Kemeny, DM; Zachary, C; et al. Oral allergy syndrome: symptoms of IgE mediated hypersensitivity to foods. *Clin. Allergy*. 1987;17:33-42.

[21] Lee, LA; Buirks, AW. Food allergies: prevalence, molecular characterization, and treatment/prevention strategies. *Annu. Rev. Nutr.* 2006;26;539-565.

[22] Ortolani, C; Ispano, M; Pastorello, EA; et al. Comparison of results of skin prick tests (with fresh foods and commercial food extracts) and RAST in 100 patients with oral allergy syndrome. *J. Allergy Clin. Immunol.* 1989;83:683-690.

[23] Hallett, R; Teuber, SS. Food allergies and sensitivities. *Nutr. Clin. Care*. 2004;7:122-129.

[24] Garcia-Careaga, M; Kerner, JA, Jr. Gastrointestinal manifestations of food allergies in pediatric patients. *Nutr. Clin. Pract.* 2005;20:526-535.

[25] Spergel, JM; Pawlowski, NA. Food allergy: mechanisms, diagnosis, and management in children. *Pediatr. Clin. North Am.* 2002;49:73-96.

[26] Strobel, S; Hourihane, OB. Gastrointestinal allergy: clinical symptoms and immunological mechanisms. *Pediatr. Allergy Immunol.* 2001;12(Suppl 14).43-46.

[27] Pastorello, EA; Robino, AM. Clinical role of lipid transfer proteins in food allergy. *Mol. Nutr. Food Res.* 2004;48:356-362.

[28] Hoffman, KM; Sampson, HA. Evaluation and management of patients with adverse food reactions. In: Bierman, CW; Pearlman, DS; Shapiro, GG; et al. (eds). *Allergy, Asthma, and Immunology from Infancy to Adulthood*. Philadelphia: WB Saunders. 1996; pp. 665-686.

[29] Burks, AW; Sampson, H. Food allergies in children. *Curr. Probl. Pediatr.* 1993;23:230-252.

[30] Sicherer, SH. Clinical aspects of gastrointestinal food allergy in childhood. *Pediatrics*. 2003;111:1609-1616.

[31] Dauer, EH; Freese, DK; El-Youssef, M; et al. Clinical characteristics of eosinophilic esophagitis in children. *Ann. Otol. Rhinol. Laryngol.* 2005;114:827-833.

[32] Furuta, GT; Straumann, A. Review article: the pathogensis and management of eosinophilic oesophagitis. *Aliment. Pharmacol. Ther.* 2006;24:173-182.

[33] Heine, RG. Pathophysiology, diagnosis and treatment of food protein-induced gastrointestinal diseases. *Curr. Opin. Allergy Clin. Immunol.* 2004;4:221-229.

[34] Spergel, JM. Eosinophilic esophagitis in adults and children: evidence for a food allergy component in many patients. *Curr. Opin. Allergy Clin. Immunol.* 2007;7:274-278.

[35] Furuta, T; Liacouras CA; Collins, MH; et al. Eosinophilic esophagitis in children and adults: a systemic review and consensus recommendations for diagnosis and treatment. *Gastroenterology.* 2007;133:1342-1363.

[36] Noel, RJ; Rothenberg, ME. Eosinophilic esophagitis. *Curr. Opin. Pediatr.* 2005;17:690-694.

[37] Martin-Muñoz, MF; Lucendo, AJ; Navarro, M, et al. Food allergies and eosinophilic esophagitis – two case studies. *Digestion.* 2006;74:49-54.

[38] Pentiuk, SP; Miller, CK; Kaul, A. Eosinophilic esophagitis in infants and toddlers. *Dysphagia.* 2007;22:44-48.

[39] Cheung, KM; Oliver, MR; Cameron, DJ, et al. Esophageal eosinophilia in children with dysphagia. *J. Pediatr. Gastroenterol. Nutr.* 2003;37:498-503.

[40] Croese, J; Fairley, SK; Masson, JW; et al. Clinical and endoscopic features of eosinophilic esophagitis in adults. *Gastrointest. Endosc.* 2003;58:516-522.

[41] Fox, VL; Nurko, S; Furuta, GT; et al. Eosinophilic esophagitis: it's not just kid's stuff. *Gastrointest. Endosc.* 2002;56:260-270.

[42] Liacouras, CA. Eosinophilic esophagitis in children and adults. *J. Pediatr. Gastroenterol. Nutr.* 2003;37:S23-S28.

[43] Remírez, JA; Escudero, R; Cáceres, O; et al. Eosinophilic esophagitis. *Allergol. Immunopathol. (Madr).* 2006;34:79-81.

[44] American College of Allergy, Asthma, & Immunology. Food allergy: a practice parameter. *Ann. Allergy Asthma Immunol.* 2006;96(Suppl 2):S1-S68.

[45] Sampson, HA; Leung, DY. Anaphylaxis. In: Kleigman, RM; Behram, RE; Jenson, HB; et al. (eds). *Nelson Textbook of Pediatrics.* Philadelphia: Saunders. 2007; pp. 983-985.

[46] Kaczmarski, SJ. Gastroesophageal reflux in children and adolescents. Clinical aspects with special respect to food hypersensitivity. *Adv. Med. Sci.* 2006;51:327-335.

[47] Kelly, KJ; Lazenby, AJ; Rowe, PC; et al. Eosinophilic esophagitis attributed to gastroesophageal reflux: improvement with an amino acid-based formula. *Gastroenterology.* 1995;109:1503-1512.

[48] Nowak-Wegrzyn, A. Food allergy to proteins. *Nestlé Nutr. Workshop Ser. Pediatr. Program.* 2007;59:17-35.

[49] Gonsalves, N. Food allergies and eosinophilic gastrointestinal illness. *Gastroenterol. Clin. North Am.* 2007;36:75-91.

[50] Klein, NC; Hargrove, RI; Sleisenger, MH; et al. Eosinophilic gastroenteritis. Medicine (Baltimore). 1970;49:299-319.

[51] Cello, JPE. Eosinophilic gastroenteritis: a complex disease entity. *Am. J. Med.* 1979;67:1097-1104.

[52] Brigino, E; Bahna, SL. Clinical features of food allergy in infants. *Clin. Rev. Allergy Immunol.* 1995;13:329-345.

[53] Proujansky, R; Winter, HS; Walker, WA. Gastrointestinal syndromes associated with food sensitivity. *Adv. Pediatr.* 1988;35:219-238.

[54] Bock, SA; Sampson, HA. Food allergy in infancy. *Pediatr. Clin. North Am.* 1994;41:1047-1067.
[55] Lake, AM. Dietary protein enterocolitis. *Curr. Allergy Rep.* 2001;1:76-79.
[56] Zapatero Remón, L; Alonso Lebrero, E; Martin Fernández, E; et al. Food protein-induced enterocolitis syndrome caused by fish. *Allergol. Immunopathol. (Madr).* 2005;33:312-316.
[57] Nowak-Wegrzyn, A; Sampson, HA; Wood, RA; et al. Food protein-induced enterocolitis syndrome caused by solid food proteins. *Pediatrics.* 2003;111:829-835.
[58] Murch, S. Allergy and intestinal dysmotility – evidence of genuine causal linkage? *Curr. Opin. Gastroenterol.* 2006;22:664-668.
[59] Yimyaem, P; Chongsrisawat, V; Vivatvakin, B, et al. Gastrointestinal manifestations of cow's milk protein allergy during the first year of life. *J. Med. Assoc. Thai.* 2003;86:116-123.
[60] Hojsak, I; Kljaić-Turkalj, M; Mišak. Z; et al. Rice protein-induced enterocolitis syndrome. *Clin. Nutr.* 2006;25:533-536.
[61] Sampson, HA. Food allergies. In: Oski, FA; DeAngelis, CD; Feigin, RD. (eds). *Principles and Practice of Pediatrics.* Philadelphia: JB Lippincott. 1994; pp. 227-232.
[62] Hirose, R; Yamada, T; Hayashida, Y. Massive bloody stools in two neonates caused by cow's milk allergy. *Pediatr. Surg. Int.* 2006;22:935-938.
[63] Swart, JF; Ultee, K. Rectal bleeding in a preterm infant as a symptom of allergic colitis. *Eur. J. Pediatr.* 2003;162:55-56.
[64] Stern, M. Gastrointestinal allergy. In: Walker, WA; Durie, PR; Hamilton, JR; et al. (eds). *Pediatric Gastrointestinal Disease.* St Louis: Mosby. 1991; pp. 557-574.
[65] Sampson, HA. Food allergies. In: McMillan, JA; Feigin, RD; DeAngelis, C; et al. *Oski's Pediatrics: Principles & Practice.* Philadelphia: Lippincott Williams & Wilkins. 2006; pp. 2417-2423.
[66] Troncone, R; Bhatnagar, S; Butzner, D; et al. Celiac disease and other immunologically mediated disorders of the gastrointestinal tract: Working Group Report of the Second World Congress of Pediatric Gastroenterology, Hepatology, and Nutrition. *J. Pediatr. Gastroenterol. Nutr.* 2004;39:S601-S610.
[67] Dewar, DH; Ciclitira, PJ. Clinical features and diagnosis of celiac disease. *Gastroenterology.* 2005;128:S19-S24.
[68] Anderson, JA. Food allergy and intolerance. In: Lieberman, P; Anderson, JA. (ed). *Current Clinical Practice: Allergic Diseases: Diagnosis and Treatment.* 3rd edition. Totowa: Humana Press, 2007, pp. 271-294.
[69] Estep, DC; Kulczycki, A, Jr. Treatment of infant colic with amino acid-based infant formula: a preliminary study. *Acta Paediatr.* 2000;89:22-27.
[70] Hill, DJ; Hudson, IL; Sheffield, LJ; et al. A low allergen diet is a significant intervention in infantile colic: results of a community-based study. *J. Allergy Clin. Immunol.* 1995;96:886-892.
[71] Iacono, G; Carroccio, A; Montalto, G; et al. Severe infantile colic and food intolerance: a long-term prospective study. *J. Pediatr. Gastroenterol. Nutr.* 1991;12:332-335.
[72] Leung, AK; Lemay, JF. Infantile colic: a review. *J. R. Soc. Health.* 2004;124:162-166.
[73] Lothe, L; Lindberg, T. Cow's milk whey protein elicits symptoms of infantile colic in colicky formula-fed infants: a double-blind crossover study. *Pediatrics.* 1989;83:262-266.

[74] Lucassen, PL; Assendelft, WJ; Gubbels, JW; et al. Infantile colic: crying time reduction with a whey hydrolysate: a double-blind, randomized, placebo-controlled trial. *Pediatrics*. 2000;106:1349-1354.
[75] Savino, F. Focus on infantile colic. *Acta Paediatr*. 2007;96:1259-1264.
[76] Hill, DJ; Hosking, CS. Infantile colic and food hypersensitivity. *J. Pediatr. Gastroenterol. Nutr.* 2000;30:S67-S76.
[77] Kalliomäki, M; Lappala, P; Korvenranta, H; et al. Extent of fussing and colic type crying preceding atopic disease. *Arch. Dis. Child.* 2001;84:349-350.
[78] Jakobsson, I; Lothe, L; Ley, D; et al. Effectiveness of casein hydrolysate feedings in infants with colic. *Acta Paediatr.* 2000; 89:18-21.
[79] Jakobbson, I; Lindberg, T. Cow's milk proteins cause infantile colic in breast-fed infants: a double-blind crossover study. *Pediatrics.* 1983;71:268-271.
[80] Hewson, P; Oberklaid, F; Menahem, S. Infant colic, distress, and crying. *Clin. Pediatr.* 1987;26:69-75.
[81] Leung, AK. Food allergy: a clinical approach. *Adv. Pediatr.* 1998;45:145-177.
[82] Nutrition Committee, Canadian Paediatric Society. Dietary manipulations for infantile colic. *Paediatr. Child Health.* 2003;8:449-452.
[83] Hill, DJ; Roy, N; Heine, RG; et al. Effect of a low-allergen maternal diet on colic among breastfed infants: a randomized, controlled trial. *Pediatrics.* 2005;116:e709-e715.
[84] Kalliomäki, MA. Food allergy and irritable bowel syndrome. *Curr. Opin. Gastroenterol.* 2005;21:708-711.
[85] Park, MI; Camilleri, M. Is there a role of food allergy in irritable bowel syndrome and functional dyspepsia? A systemic review. *Neurogastroenterol. Motil.* 2006;18:595-607.
[86] Gonsalkorale, WM; Perrey, C; Pravica, V; et al. Interleukin 10 genotypes in irritable bowel syndrome: evidence for an inflammatory component? *Gut.* 2003;52:91-93.
[87] van der Veek, PP; van den Berg, M; de Kroon, YE; et al. Role of tumor necrosis factor-α and interleukin-10 gene polymorphisms in irritable bowel syndrome. *Am. J. Gastroenterol.* 2005;100:2510-2516.
[88] Barbara, G; Stanghellini, V; de Giorgio, R; et al. Activated mast cells in proximity to colonic nerves correlate with abdominal pain in irritable bowel syndrome. *Gastroenterology.* 2004;126:693-702.
[89] Atkinson, W; Sheldon, TA; Shaath, N; et al. Food elimination based on IgG antibodies in irritable bowel syndrome: a randomized controlled trial. *Gut.* 2004;53:1459-1464.
[90] Isolauri, E; Rautava, S; Kalliomäki, M. Food allergy in irritable bowel syndrome: new facts and old fallacies. *Gut.* 2004;53:1391-1393.
[91] Leung, AK; Wong, BE; Cho, HY, et al. Recurrent abdominal pain in childhood. *Singapore Pediatr. J.* 1996;38:44-48.
[92] Leung, AK; Lemay, JF; Barker, C. Recurrent abdominal pain in children. *Can. J. Diagn.* 2002;19(5):68-80.
[93] Kokkonen, J; Ruuska, T; Karttunen, T; et al. Mucosal pathology of the foregut associating with food allergy and recurrent abdominal pains in children. *Acta Paediatr.* 2001;90:16-21.
[94] Husby, S; Host, A. Recurrent abdominal pain, food allergy and endoscopy. *Acta Paediatr.* 2001;90:3-4.

[95] Leung, AK; Chan, PY; Cho, HY. Constipation in children. *Am. Fam. Physician.* 1996;54:611-627.
[96] Loening-Baucke, V. Prevalence, symptoms and outcome of constipation in infants and toddlers. *J. Pediatr.* 2005;146:359-363.
[97] Iacono, G; Carroccio, A; Cavataio, F; et al. Chronic constipation as a symptom of cow milk allergy. *J. Pediatr.* 1995;126:34-39.
[98] Iacono, G; Cavataio, F; Montalto,G; et al. Intolerance of cow's milk and chronic constipation in children. *N. Engl. J. Med.* 1998;339:1100-1104.
[99] Daher, S; Tahan, S; Solé, D; et al. Cow's milk protein intolerance and chronic constipation in children. *Pediatr. Allergy Immunol.* 2001;1:339-342.
[100] Carroccio, A; Scalici, C; Maresi, E; et al. Chronic constipation and food intolerance: a model of proctitis causing constipation. *Scand. J. Gastroenterol.* 2005;40:33-42.
[101] Burks, W. Skin manifestations of food allergy. *Pediatrics.* 2003;111:1617-1624.
[102] Fasano, MB. Dermatologic food allergy. *Pediatr. Ann.* 2006;35:727-731.
[103] Host, A; Halken, S. A prospective study of cow milk allergy in Danish infants during the first 3 years of life. *Allergy.* 1990;45:587-596.
[104] Hill, DJ; Firer, MA; Shelton, MJ; et al. Manifestations of milk allergy in infants: clinical and immunologic findings. *J. Pediatr.* 1986;109:270-276.
[105] Champion, RH; Roberts, SO; Carpenter RG; et al: Urticaria and angioedema: a review of 554 patients. *Br. J. Dermatol.* 1969;81:588-597.
[106] Lemanske, RF Jr; Sampson, HA. Adverse reactions to foods and their relationships to skin diseases in children. *Adv. Pediatr.* 1988;35: 89-218.
[107] Winston, GB; Lewis, CW. Contract dermatitis. *Int. J. Dermatol.* 1982;21:573-578.
[108] Pastar, Z; Lipozencic, J. Adverse reactions to food and clinical expressions of food allergy. *SKINmed.* 2006;5:119-125.
[109] Sampson, HA. The immunopathogenic role of food hypersensitivity in atopic dermatitis. *Acta Derm. Venereol.* 1992;176:S34-S37.
[110] Gontzes, P; Bahna, SL. Food allergy for the primary care physician. *Primary Care.* 1987;14:547-558.
[111] Leung, AK; Hon, KL; Robson, WL. Atopic dermatitis. *Adv. Pediatr.* 2007;54:241-273.
[112] Burks, AW; Mallroy, SB; Williams, LW; et al. Atopic dermatitis: clinical relevance of food hypersensitivity reactions. *J. Pediatr.* 1988;113:447-451.
[113] Sampson, HA; McCaskill, CC. Food hypersensitivity and atopic dermatitis: evaluation of 113 patients. *J. Pediatr.* 1985;107:669-675.
[114] Sampson, HA; Mendelson, L; Rosen, JP. Fatal and near-fatal anaphylactic reactions to food in children and adolescents. *N. Engl. J. Med.* 1992;327:380-384.
[115] Breuer, K; Heratizadeh, A; Wulf, A; et al. Late eczematous reactions to food in children with atopic dermatitis. *Clin. Exp. Allergy.* 2004;34:817-824.
[116] Hill, DJ; Hosking, CS. Food allergy and atopic dermatitis in infancy: an epidemiology study. *Pediatr. Allergy Immunol.* 2004;15:421-427.
[117] Leung, AK; Bowen, TJ. Seasonal allergic rhinitis and food allergy. In: Bergman, AB. (ed). *Twenty Common Problems in Pediatrics.* New York: McGraw-Hill, 2001, pp. 219-233.
[118] Dolovich, J; Hargreave, FE; Chalmers, R; et al. Late cutaneous allergic responses in isolated IgE-dependent reactions. *J. Allergy Clin. Immunol.* 1973;52:38-46.

[119] Solley, GO; Gleich, GJ; Jordan, RE; et al. The late phase of the immediate wheal and flare skin reactions: its dependence on IgE antibodies. *J. Clin. Invest.* 1976;58:408-420.

[120] Lio, PA. Atopic dermatitis and food allergies: true, true and related? *Arch. Dis. Child. Educ. Pract. Ed.* 2007;92:ep56-ep60.

[121] Hill, DJ; Duke, AM; Hosking, CS; et al. Clinical manifestations of cow's milk allergy in childhood. II. The diagnostic value of skin tests and RAST. *Clin. Allergy.* 1988;18:481-490.

[122] Brancaccio, RR; Alvarez, MS. Contact allergy to food. *Dermatol. Ther.* 2004;17:302-313.

[123] Judd, L. A descriptive study of occupational skin disease. *N. Z. Med. J.* 1994;107: 147-149.

[124] Sampson, HA. Update on food allergy. *J. Allergy Clin. Immunol.* 2004;113:805-819.

[125] Ekeowa-Anderson, AL. Allergic contact cheilitis to garlic. *Contact Dermatitis.* 2007;56:174-175.

[126] Tamagawa-Mineoka, R; Katoh, N; Kishimoto, S. Allergic contact cheilitis due to geraniol in food. *Contact Dermatitis.* 2007;56:242-243.

[127] Hall, RP. Dermatitis herpetiformis. *J. Invest. Dermatol.* 1992;99:873-881.

[128] Solheim, BG; Ek, J; Thune, PO; et al. HLA antigens in dermatitis herpetiformis and coeliac disease. *Tissue Antigens.* 1976;7:57-59.

[129] El-Gamal, YM; Hossny, EM. Respiratory food allergy. *Pediatr. Ann.* 2006;35:733-740.

[130] James, JM. Respiratory manifestations of food allergy. *Pediatrics.* 2003;111:1625-1630.

[131] Ortolani, C. Atlas on mechanisms in adverse reactions to food. *Allergy.* 1995;50:5-81.

[132] James, JM; Crespo, JF. Allergic reactions to foods by inhalation. *Curr. Allergy Asthma Rep.* 2007;7:167-174.

[133] Ozol, D; Mete, E. Asthma and food allergy. *Curr. Opin. Pulm. Med.* 2008;14:9-12.

[134] Onorato, J; Merland, N; Terral, C; et al: Placebo-controlled double-blind food challenge in asthma. *J. Allergy Clin. Immunol.* 1986;78:1139-1146.

[135] Roberts, G; Lack, G. Food allergy and asthma – what is the link? *Paediatr. Respir. Rev.* 2003;4:205-212.

[136] Rance, F; Micheau, P; Marchac, V; et al. Food allergy and asthma in children. *Rev. Pneumol. Clin.* 2003;59:109-113.

[137] James, JM; Eggleston, PA; Sampson, HA. Food allergy increases airway reactivity. *Am. J. Crit. Care Respir. Med.* 1994;149:59-64.

[138] Roberts, G; Lack, G. Relevance of inhalational exposure to food allergens. *Curr. Opin. Allergy Clin. Immunol.* 2003;3:211-215.

[139] Crespo, JF; Pascual, C; Dominguez, C; et al. Allergic reactions associated with airborne fish particles in IgE-mediated fish hypersensitive patients. *Allergy.* 1995; 50:257-261.

[140] Roberts, G; Golder, N; Lack, G. Bronchial challenges with aerosolized food in asthmatic, food-allergic children. *Allergy.* 2002;57:713-717.

[141] Sicherer, SH; Furlong, TJ; DeSimone, J; et al. Self-reported allergic reactions to peanuts on commercial airlines. *J. Allergy Clin. Immunol.* 1999;104:186-189.

[142] Simpson, AB; Glutting, J; Yousef, E. Food allergy and asthma morbidity in children. *Pediatr. Pulmonol.* 2007;42:489-495.

[143] Wang, J; Visness, CM; Sampson, HA. Food allergen sensitization in inner-city children with asthma. *J. Allergy. Clin. Immunol.* 2005;115:1076-1080.

[144] Gustafsson, D; Sjöberg, O; Foucard, T. Sensitization to food and airborne allergens in children with atopic dermatitis followed up to 7 years of age. *Pediatr. Allergy Immunol.* 2003;14:448-452.

[145] Kotaniemi-Syrjanen, A; Reijonen, T; Romppanen, J. Allergen-specific immunoglobulin E antibodies in wheezing infants: the risk for asthma in later childhood. *Pediatrics.* 2003;111:e255-e261.

[146] Heiner, DC; Sears, JW. Chronic respiratory disease associated with multiple circulating precipitins to cow's milk. *Am. J. Dis. Child.* 1960;100:500-502.

[147] Bahna, SL. Adverse food reactions by skin contact. *Allergy.* 2004;59(Suppl 78):66-70.

[148] Aydoğan, B; Kiroğlu, M; Altintas, D; et al. The role of food allergy in otitis media with effusion. *Otolaryngol. Head Neck Surg.* 2004;130:747-750.

[149] Bernstein, JM; Lee, J; Conboy, K; et al. Further observations on the role of IgE-mediated hypersensitivity in recurrent otitis media with effusion. *Otolaryngol. Head Neck Surg.* 1985;93:611-615.

[150] Luyasu, S; Morisset, M; Guenard, L; et al. Acute recurrent otalgia and food allergy: a case report and review of the literature. *Eur. Ann. Allergy Clin. Immunol.* 2005;37:60-62.

[151] Nsouli, TM; Nsouli, SM; Linde, RE; et al. Role of food allergy in serous otitis media. *Ann. Allergy.* 1994;73:215-219.

[152] Juntti, H; Tikkanen, S; Kokkonen, J; et al. Cow's milk allergy is associated with recurrent otitis media during childhood. *Acta Otolaryngol.* 1999;119:867-873.

[153] Derebery, MJ; Berliner, KI. Prevalence of allergy in Meniere's disease. *Otolaryngol. Head Neck Surg.* 2000;123:69-75.

[154] Keleş, E; Gödekmerdan, A; Kalidağ, T; et al. Ménière's disease and allergy: allergens and cytokines. *J. Laryngol. Otol.* 2004;118:688-693.

[155] Golbert, TM; Patterson, R; Pruzansky, JJ. Systemic allergic reactions to ingested antigens. *J. Allergy.* 1969;44:96-107.

[156] Joint Task Force on Practice Parameters; American Academy of Allergy, Asthma and Immunology; American College of Allergy, Asthma and Immunology; Joint Council of Allergy, Asthma and Immunology. The diagnosis and management of anaphylaxis: an updated practice parameter. *J. Allergy Clin. Immunol.* 2005;115:S483-S523.

[157] Sampson, HA. Anaphylaxis and emergency treatment. *Pediatrics.* 2003;111: 1601-1608.

[158] Brown, SG. Cardiovascular aspects of anaphylaxis: implications for treatment and diagnosis. *Curr. Opin. Allergy Clin. Immunol.* 2005;5:359-364.

[159] Wang, J; Sampson, HA. Food anaphylaxis. *Clin. Exp. Allergy.* 2007;37:651-660.

[160] Maulitz, RM; Pratt, DS; Schocket, AL. Exercise-induced anaphylactic reactions to shellfish. *J. Allergy Clin. Immunol.* 1979;63:433-434.

[161] Sheffer, AL; Austen, KF. Exercise-induced anaphylaxis. *J. Allergy Clin. Immunol.* 1980;66:106-111.

[162] Kidd, JM III; Cohen, SH; Sosman, AJ; et al. Food-dependent exercise-induced anaphylaxis. *J. Allergy Clin. Immunol.* 1983;71:407-411.

[163] Beaudouin, E; Renaudin, JM; Morisset, M; et al. Food-dependent exercise-induced anaphylaxis – update and current data. *Eur. Ann. Allergy Clin. Immunol.* 2006;38:45-51.
[164] Castells, MC; Horan, RF; Sheffer, AL. Exercise-induced anaphylaxis (EIA). *Clin. Rev. Allergy Immunol.* 1999;17:413-424.
[165] Noma, T; Yoshizawa, I; Ogawa, N; et al. Fatal buckwheat dependent exercise induced anaphylaxis. *Asian Pacific J. Allergy Immunol.* 2001;19:283-286.
[166] Senna, G; Mistrello, G; Roncarolo, D; et al. Exercise-induced anaphylaxis to grape. *Allergy.* 2001;56:1235.
[167] Dohi, M; Suko, M; Sugiyama, H; et al. 3 cases of food-dependent exercise-induced anaphylaxis in which aspirin intake exacerbated anaphylactic symptoms. *Arerugi.* 1990;39:1598-1604.
[168] Morita, E; Kunie, K; Matsuo, H. Food-dependent exercise-induced anaphylaxis. *J. Dermatol. Sci.* 2007;47:109-117.
[169] Anderson, JA. The clinical spectrum of food allergy in adults. *Clin. Exp. Allergy.* 1991;21:S304-S315.
[170] Silverstein, SR; Frommer, DA; Dobozin, B; et al. Celery-dependent exercise-induced anaphylaxis. *J. Emerg. Med.* 1986;4:195-199.
[171] Buchbinder, EM; Bloch, KJ; Moss, J; et al. Food-dependent, exercise-induced anaphylaxis. *JAMA.* 1983;250:2973-2974.
[172] Feingold, BF. Food additives and child development. *Hosp. Pract.* 1973;8:11-21.
[173] Feingold, BF. Hyperkinesis and learning disabilities linked to artificial food flavors and colors. *Am. J. Nurs.* 1975;75:797-803.
[174] Leung, AK; Robson, WL; Fagan, JE; et al. Attention-deficit hyperactivity disorder: getting control of impulsive behavior. *Postgrad. Med.* 1994;95:153-160.
[175] Leung, AK; Lemay, JF. Attention deficit hyperactivity disorder: an update. *Adv. Ther.* 2003;20:303-318.
[176] Varley, CK. Diet and the behavior of children with attention deficit disorder. *J. Am. Acad. Child Psychiatry.* 1984;23:182-185.
[177] Monro, J; Carini, C; Brostoff, J. Migraine is a food-allergic disease. *Lancet.* 1984;1:719-721.
[178] Mansfield, LE; Vaughan, TR; Waller, SF; et al. Food allergy and adult migraine: double-blind and mediator confirmation of an allergic etiology. *Ann. Allergy.* 1985;55:126-129.
[179] Egger, J; Carter, CM; Soothill, JF; et al. Oligoantigenic diet treatment of children with epilepsy and migraine. *J. Pediatr.* 1989;114:51-58.
[180] Fotherby, KJ; Hunter, JO. Symptoms of food allergy. *Clin. Gastroenterol.* 1985;14:615-629.
[181] Pearson, DJ; McKee, A. Food allergy. *Adv. Nutr. Res.* 1985;7:1-37.
[182] Johansson, SG; Hourihane, JO; Bousquet, J; et al. A revised nomenclature for allergy. An EAACI position statement from the EAACI nomenclature task force. *Allergy.* 2001;56:813-824.
[183] Crayton, JW; Stone, T; Stein, G. Epilepsy precipitated by food sensitivity: report of a case with double-blind placebo-controlled assessment. *Clin. Electroencephalographol.* 1981;12:192-198.

[184] Frediani, T; Lucarelli, S; Pelliccia, A; et al. Allergy and childhood epilepsy: a close relationship? *Acta Neurol. Scand.* 2001;104:349-352.

[185] Frediani, T; Pelliccia, A; Aprile, A; et al. Partial idiopathic epilepsy: recovery after allergen-free diet. *Pediatr. Med. Chir.* 2004;26:196-197.

[186] Cavallo, MG; Fava, D; Monetini, L; et al. Cell-mediated immune response to ß casein in recent-onset insulin-dependent diabetes: implications for disease pathogenesis. *Lancet.* 1996;348:926-928.

[187] Harrison, LC. Cow's milk and IDDM. *Lancet.* 1996;348:905-906.

[188] Karjalainen, J; Martin, JM; Knip, M; et al. A bovine albumin peptide as a possible trigger of insulin-dependent diabetes mellitus. *N. Engl. J. Med.* 1992;327:302-307.

[189] Norris, JM; Beaty, B; Klingensmith, G et al. Lack of association between early exposure to cow's milk protein and ß-cell autoimmunity: diabetes autoimmunity study in the young (DAISY). *JAMA.* 1996;276:609-614.

[190] Sandberg, DH; McIntosh, RM; Bernstein, CW; et al. Severe steroid-responsive nephrosis associated with hypersensitivity. *Lancet.* 1997;1:388-391.

[191] Genova, R; Sanfilippo, M; Rossi, ME; et al. Food allergy in steroid-resistant nephrotic syndrome. *Lancet.* 1987;1:1315-1316.

[192] Laurent, J; Rostoker, G; Robera, R; et al. Is adult idiopathic nephrotic syndrome food allergy? Value of oligoantigenic diets. *Nephron.* 1987;47:7-11.

[193] Sieniawska, M; Szymanik-Grzelak, H; Kowalewska, M; et al. The role of cow's milk protein intolerance in steroid-resistant nephrotic syndrome. *Acta Paediatr.* 1992;81:1007-1012.

[194] Mungan, NA; Seckiner, I; Yesilli, C; et al. Nocturnal enuresis and allergy. *Scand. J. Urol. Nephrol.* 2005;39:237-241.

[195] Oei, HD; Pelikan-Filipek, M; Pelikan, Z; et al. Enuresis and encopresis as a reaction to food. *Ned. Tijdschr. Geneeskd.* 1989;133:1555-1557.

[196] Rawashdeh, YF; Hvistendahl, GM; Kamperis, K; et al. Demographics of enuresis patients attending a referral centre. *Scand. J. Urol. Nephrol.* 2002;36:348-353.

[197] Whitefield, MF; Barr, DGD. Cow's milk allergy in the syndrome of thrombocytopenia with absent radius. *Arch. Dis. Child.* 1976;51:337-343.

[198] Jones, RHT. Congenital thrombocytopenia and milk allergy. *Arch. Dis. Child.* 1977;52:744-745.

[199] Caffrey, EA; Sladen, GE; Isaacs, PET; et al. Thrombocytopenia caused by cow's milk. *Lancet.* 1981;2:316.

[200] Businco, L; Falconieri, P; Bellioni-Businco, B; et al. Severe food-induced vasculitis in two children. *Pediatr. Allergy Immunol.* 2002;12:68-71.

[201] Lunardi, C; Bambara, LM; Biasi, D; et al. Elimination diet in the treatment of selected patients with hypersensitivity vasculitis. *Clin. Exp .Rheum.* 1992;10:131-135.

[202] Panush, RS; Carter, RL; Katz, P; et al. Diet therapy for rheumatoid arthritis. *Arthritis Rheum.* 1983;26:462-471.

[203] Panush, RS; Stroud, RM; Webster, EM. Food-induced (allergic) arthritis: Inflammatory arthritis exacerbated by milk. *Arthritis Rheum.* 1986;29:220-226.

[204] Panush, RS. Food induced ("allergic") arthritis: clinical and serologic studies. *J. Rheumatol.* 1990;17:291-294.

[205] Parke, AL; Hughes, GR. Rheumatoid arthritis and food: a case study. *Br. Med. J.* 1981;282:2027-2029.

[206] Golding, DN. Is there an allergic synovitis? *J. R. Soc. Med.* 1990;83:312-314.

[207] van der Laar, MA; van der Korst, JK. Food intolerance in rheumatoid arthritis. I. A double blind, controlled trial of the clinical effects of elimination of milk allergens and azo dyes. *Ann. Rheum. Dis.* 1992;51:298-302.

[208] Denman, AM; Mitchell, B; Ansell, BM. Joint complaints and food allergic disorders. *Ann. Allergy.* 1983;51:260-263.

[209] Karatay, S; Erdem, T; Yildirim, K; et al. The effect of individualized diet challenges consisting of allergenic foods on TNF-α and IL-1β levels in patients with rheumatoid arthritis. *Rheumatology.* 2004;43:1429-1433.

[210] Bahna, SL. Unusual presentations of food allergy. *Ann. Allergy Asthma Immunol.* 2001;86:414-420.

In: Food Allergies: New Research
Editor: Carrie M. Chesterton, pp. 121-133

ISBN 978-1-60456-978-0

Chapter 3

FOOD ALLERGIES AND ATOPIES: WHAT IS THE EVIDENCE?

***Kam-lun Ellis Hon*[1] *and Alexander K.C. Leung*[2,*]**

[1]Chinese University of Hong Kong,
Prince of Wales Hospital, Shatin, Hong Kong
[2]University of Calgary,
Alberta Children's Hospital, Calgary, Alberta, T2M 0H5, Canada

ABSTRACT

Genuine food allergy affects approximately 5% of children and less than 1% of adults. The underlying mechanism is complex and involves immediate (type I) or delayed (type IV) sensitization to food proteins. The gold standard for the diagnosis of genuine food allergies is by double-blind placebo-controlled food challenge test. The literature gives ambiguous data on the association between food allergies and atopic diseases. In asthma, aeroallergens such as house dust, mites, and pollens are well known allergens. Apart from type I hypersensitivity reaction precipitating acute anaphylactic and asthmatic attacks by peanuts, egg or crustacean seafood, the association with food allergens is probably less prevalent. Allergic rhinitis/allergic rhinoconjunctivitis as the sole manifestation of food allergy is quite uncommon. Food allergy plays an immunopathogenic role in 30 to 50% of children with moderate to severe atopic dermatitis.

Keywords: food allergy, asthma, allergic rhinitis, atopic dermatitis

* Correspondence to:Dr. Alexander K.C. Leung, #200, 233-16th Avenue NW, Calgary, Alberta, Canada T2M 0H5; Telefax: (403) 230-3322, E-mail: aleung@ucalgary.ca

Incidence of Food Allergy

Food allergy is a very popular topic in childhood atopies. The exact incidence of food allergy varies between study populations. Up to 25% of adults believe that they or their children are afflicted with a food allergy [1]. However, the actual prevalence of food allergy in children during their first 3 years of life is approximately 6 to 8%, and many children then develop clinical tolerance. Genuine food allergy affects approximately 5% of children and less than 1% of adults in various studies [2]. Food allergy encompasses a whole spectrum of disorders, with symptoms that may be cutaneous, gastrointestinal or respiratory in nature. The most common food allergens in adults are shellfish, peanuts, tree nuts, fish and eggs, and the most common food allergens in children are milk, eggs, peanuts, and tree nuts [1-5]. The definition of food allergy is, however, often quite imprecise. Venter et al. followed a birth cohort of 966 infants on the Island of Wight, United Kingdom born between September 2001 and August 2002 to the age of one year [6]. Cumulative incidence of parentally reported food hypersensitivity was 25.8%. Open or double-blind, placebo-controlled food challenges were used to confirm suspected reactions. Only 2.2% of those tested were confirmed to have food allergy, indicating the need to evaluate suspected food allergy to avoid needless dietary restriction. Food allergy is an abnormal immunological reaction to a food or food additive that causes an adverse reaction. Food allergy is therefore distinct from other adverse responses to food which may share similar clinical features, such as food intolerance, pharmacologic reactions, and toxin-mediated reactions.

Pathophysiology of Food Allergy

The underlying mechanism of food allergy is complex and involves immediate (type I) or delayed (type IV) sensitization to food proteins [1,7,8]. The immune pathogenesis is in the majority of cases, IgE-mediated although it may also be cell-mediated (non-IgE) or mixed IgE/cell-mediated. Generally, introduction of allergens through the digestive tract is thought to induce immune tolerance. In atopic individuals, the immune system produces IgE antibodies against protein epitopes on non-pathogenic substances, including dietary components. The IgE molecules are coated onto mast cells, which inhabit the mucosal lining of the digestive tract. Upon ingesting an allergen, the IgE reacts with its protein epitopes and degranulates the mast cells to release a number of chemicals (including histamine). Immediate acute reaction usually occurs within seconds to 1 to 2 hours following consumption of the food allergen and may include angioedema of the eyelids, face, lips, tongue, and trachea. There may be itching of the mouth, throat, eyes, or skin. Gastrointestinal symptomatology includes nausea, vomiting, diarrhea, and abdominal cramps. Upper airway symptoms include rhinorrhea and nasal congestion, whereas lower airway symptoms include wheezing and shortness of breath. Severe anaphylaxis may lead to hypotension, loss of consciousness, and even death. In the delay type of food allergy, the gastrointeestinal tract (in the form of esophagitis, gastroenteritis, colitis, enteropathy) and skin (in the form of eczema) are often involved. Nevertheless, the term "food allergy" is loosely used by layman, among parents and by the press, and may encompass such conditions as lactose and fructose intolerance which has nothing to do with food allergies.

Epidemiological trends of allergic diseases and asthma in children reveal a global rise in their prevalence over the past 50 years. As an explanation for the increasing trends of atopies, the "hygiene theory" has become popular in recent years. The theory speculates that in the modern, industrialized nations, food allergy is more common due to the lack of early exposure to dirt and germs, in part due to the overuse of antibiotics and antibiotic cleansers. This "over-simplistic" theory is based partly on studies showing low incidence allergy in third world countries. The body, with less dirt and germs to fight off (the T-helper 1 immune system), turns on itself and attacks food proteins as if they were foreign invaders (the T-helper 2 system) [9-12].

Diagnosing Food Allergy

A detailed dietary history or diary is of utmost importance. Various tests may be performed if the symptoms are consistent with allergy to a particular food. The gold standard for the diagnosis of genuine food allergies is by double-blind placebo-controlled food challenge test. The procedure requires tertiary and dietitian supports. It is tedious, cumbersome and expensive, and is seldom feasible in general practice.

Skin prick testing is easy to do and results are available in minutes. In these tests, a tiny amount of the suspected allergen is introduced into the epidermis by a prick. This puts a small amount of the allergen under the skin. A hive will form if the person is allergic to the allergen. The positive predictive values are generally low [13]. Skin tests cannot predict if a reaction would occur or what kind of reaction might occur if a person ingests that particular allergen. Nevertheless, they can confirm an allergy in light of a patient's history of reactions to a particular food. Negative skin tests essentially confirm the absence of IgE-mediated allergic reactivity (negative predictive accuracy >95%) [1]. Non-IgE mediated allergies, which are probably more relevant in eczema, cannot be detected by this method. Skin-prick testing is often used to identify food sensitization, although double-blind placebo-controlled food challenge tests remain the gold standard for diagnosis. Recent evidence suggests that quantitative IgE measurements can predict the outcome of double-blind placebo-controlled food challenge tests and can replace about half of all oral food challenges. When a meticulous medical history is obtained in combination with IgE quantification, even fewer patients may require formal food challenges. It has also become possible to map the IgE-binding regions of many major food allergens. This may help to identify children with persistent food allergy, as opposed to those who may develop clinical tolerance. In future, microarray technology may enable physicians to screen patients for a large number of food proteins and epitopes, using just a few drops of blood [1].

Blood tests are another useful diagnostic tool for evaluating IgE-mediated food allergies. The RAST (RadioAllergoSorbent Test) detects the presence of IgE antibodies to a particular allergen. A CAP-RAST test is a specific type of RAST test with greater specificity: it can show the amount of IgE present to each allergen. Predictive values for certain foods have been determined [14]. These predictive values can be compared to the RAST blood test results. If the RAST score is higher than the predictive value for that food, then there is over a 95% chance that the person will have an allergic reaction (limited to rash and anaphylaxis reactions) if the food is ingested. Currently, predictive values are available for the following foods: milk, egg, peanut, fish, soy, and wheat [15]. Blood tests allow for hundreds of allergens to be screened from a single sample, and cover food allergies as well as inhalants. However, non-IgE mediated allergies cannot be detected by this method.

Blood testing methodologies currently available that can measure IgG are not acceptable as a method of allergy evaluation as IgG antibodies are not implicated in a food allergy reaction. The significance of IgG anti-allergen antibodies was lacking as reviewed by the American Academy of Allergy and Immunology. Although a number of commercial labs sell tests that reportedly measure IgG antibodies against common allergens, there is no clinical significance of such findings. It is not established that these commercial assays actually measure the IgG antibodies that they report. Also, even if the assays are measuring IgG anti-allergen antibodies, the clinical significance of such antibodies is certainly not established. The significance of IgG anti-food antibodies is particularly questionable since many children with such antibodies in their serum tolerate the foods in question perfectly well. There has been no study to date to validate the usefulness of IgG.

In recent years, the atopy patch test (APT) employed in conjunction with specific IgE assays or skin prick test have been found to be helpful in the diagnosis and treatment of patients with food allergy [7]. The atopy patch test for each food allergen has to be validated with double-blind, placebo-controlled food challenge. There is a need for larger studies in order to investigate what the best formulations might be for atopy patch tests. It has been demonstrated that atopy patch tests, together with determination of specific IgE levels, reduce the need for oral food challenges in children with atopic dermatitis [15].

One study investigated 437 children (median age, 13 months; 90% with atopic dermatitis) referred for evaluation of suspected food allergy. Specific serum IgE measurements, skin prick tests, atopy patch tests, and placebo-controlled oral food challenges were performed. Eight hundred and seventy three oral challenges with cow's milk, hen's egg, wheat, and/or soy were analyzed. One thousand seven hundred single atopy patch tests were performed. As a single parameter, the atopy patch tests showed the best specificity compared with specific serum IgE measurements, skin prick tests, or both. Combining the atopy patch test with either the skin prick test or specific serum IgE measurement resulted in improved sensitivity and specificity. Decision points for specific serum IgE measurement and for the skin prick test showed lower values when combined with a positive atopy patch test result. With combined testing, including atopy patch tests, only between 0.5% and 7% (99% predicted probability) and between 6% and 14% (using 95% predicted probability) of children would fulfill the criteria for avoiding an oral food challenge. The predictive capacity of the atopy patch test is improved when combined with specific serum IgE measurement or the skin prck test; oral food challenges become superfluous in only 0.5% to 14% of study patients. The study shows that the atopy patch test is time-consuming and demands a highly experienced test evaluator. For daily clinical practice, the atopy patch test adds only a small predictive value to the standard skin prick test and specific serum IgE measurement in the diagnostic workup of suspected food-related symptoms in our study population [16]. As the sensitivity of atopy patch test is generally low, the test must be standardized against double-blind placebo-controlled food challenge test [17,18].

The interpretation of the atopy patch test to many foods has not been standardized. The study by Heine et al. aimed to validate the reading of the atopy patch test in terms of the diagnostic accuracy of individual skin signs [18]. Eighty-seven children (mean age 2.4 ± 2.5 yr, range 0.5-13.5; 57 male) with atopic dermatitis and suspected food allergies underwent atopy patch test to cow's milk, hen's egg, wheat and soy. Twelve-millimetre Finn chambers were applied for 48 hours, and results were read after 48 and 72 hours. Skin changes were graded for erythema, induration, papule formation and "crescendo" phenomenon (increase of skin sign severity from 48 to 72 hours). Food allergy was assessed by double-blind, placebo-controlled food challenges. Sensitivity, specificity and predictive values were calculated for each skin signs in relation to challenge outcome. Of 165 double-blind, placebo-controlled food challenges, 75 (45%) were positive. The combination of any skin induration plus papules (seven or more), or of moderate erythema plus any induration plus seven or more papules had a positive predictive value and specificity for the challenge outcome of 100%; however, the sensitivity was low (8% to 15%). The best diagnostic accuracy for single signs was found for induration beyond the Finn chamber margin (positive predictive value 88%, specificity 99%, sensitivity 9%) and presence of at least seven papules (positive predictive value 80%, specificity 96% sensitivity 21%). Presence of both induration and at least seven papules at 72 hours were the skin signs with the greatest diagnostic accuracy for food allergy in children with atopic dermatitis [18].

Food challenges, especially double-blind placebo-controlled food challenges, are the gold standard for diagnosis of food allergies, including most non-IgE mediated reactions [1,3,7,8]. Blind food challenges involve packaging the suspected allergen into a capsule, giving it to the patient, and observing the patient for signs or symptoms of an allergic reaction. Due to the risk of anaphylaxis, food challenges are usually conducted in a hospital environment in the presence of a doctor.

Additional diagnostic tools for evaluation of eosinophilic or non-IgE mediated reactions include endoscopy, colonscopy and biopsy. These tests are occasionally performed in less strict forward cases.

FOOD ALLERGY AND ATOPIES: THE ASSOCIATION

The literature gives ambiguous data on the association between food allergies and atopic diseases. Apart from acute severe reactions such as anaphylaxis, it appears that food allergy plays a more significant role in young children with eczema but not airway atopies such as allergic rhinoconjunctivitis or asthma. The concept that children at risk of developing allergic diseases follow an 'atopic march' in that these children manifest food allergy and eczema during infancy which 'march' later into airway allergies of asthma and allergic rhinitis has recently be challenged [12]. Evidence suggests that the risk of subsequent childhood asthma is not increased in children with early atopic dermatitis who are not also early wheezers, suggesting a co-manifestation of phenotypes rather than a progressive atopic march [12].

Food Allergy and Asthma

In asthma, aeroallergens such as house dust, mites, and pollens are well known allergens. Apart from type I hypersensitivity reaction precipitating acute anaphylactic and asthmatic attacks by peanuts, egg or crustacean seafood, the association with food allergens is probably less prevalent. Food-triggered asthma is rare and estimated to occur only among 6 to 8% of children with asthma and less than 2% of adults with asthma. Contrary to the beliefs of many, there are only very few confirmed food triggers of asthma. Milk, eggs, peanuts, tree nuts, soy, wheat, fish, shellfish and sulphites and sulphiting agents in foods commonly found in dried fruits have been found to trigger asthma [19,20]. Many food ingredients such as food dyes and colors, food preservatives (butylated hydroxyanisole (BHA) and the related compound butylated hydroxytoluene (BHT) are phenolic compounds that are often added to foods to preserve fats), monosodium glutamate, aspartame, and nitrite have not been conclusively linked to asthma.

There has been no study in children but one study in adults to evaluate the relationship between food allergy and asthma morbidity. A cohort of persistent asthmatics from an inner-city clinic was interviewed [20]. Allergies to food were assessed by patient report of convincing symptoms of acute allergic reactions. Outcome variables included health resource utilization and medication use. The prevalence of allergy to fish, peanut, tree-nut, shellfish, and seed allergies were 3%, 3%, 3%, 13%, and 1%, respectively. Patients with allergies to more than one food had increased incidence of asthma hospitalizations, emergency department visits, and use of oral steroids ($p < 0.05$ for all comparisons). Specifically, allergy to fish was associated with a greater risk of health resource utilization and increased frequency of oral steroid use ($p \leq 0.03$ for all comparisons). Self-reported allergy to foods was associated with worse outcomes, suggesting that food allergy may be a risk factor for increased asthma morbidity in adults [20]. Another study sought to determine those factors present in early life that predict an increased risk of adult asthma [19]. A prospective cohort study of subjects at risk of asthma and atopy was undertaken in Poole, England. One hundred babies of atopic parents were recruited at birth. During the first 5 years of life, subjects were recalled annually, all respiratory events were reported, and skin prick tests and total serum IgE measurements were performed. At 11 and 22 years, bronchial hyperresponsiveness was also measured. Seventy-three subjects were followed up at 5 years, 67 at 11 years, and 63 at 22 years. Twenty-three (37%) adult subjects reported wheezing within the previous 12 months. Fifteen (25%) of these subjects showed signs of bronchial hyperresponsiveness and were regarded as asthmatic. Wheezing before the age of 2 years occurred in 28% and was not significantly related to adult asthma (odds ratio: 0.3; 95% confidence interval: 0.03 to 1.7; $P = .19$). A positive skin prick test response to hen's egg, cow's milk, or both in the first year was independently predictive of adult asthma (odds ratio: 10.7; 95% confidence interval: 2.1 to 55.1; $P = .001$; sensitivity: 57%; specificity: 89%). The authors conclude that prediction of adult asthma remains difficult. In this study of subjects at risk of atopy, skin sensitivity to hen's egg or cow's milk in the first year was predictive of adult asthma [19].

In the meta-analysis performed by Ram et al (six trials), infants that were fed hydrolyzed formula for at least four months had reduced risk of asthma or wheeze during the first year of life compared to those fed standard cow's milk-based formula (relative risk: 0.40; 95% confidence intervals 0.19 to 0.85) [21]. Feeding soya-based formula as opposed to standard cow's milk formula did not reduce the risk of having asthma or wheeze at any age. The authors recommend that breast-milk should remain the food of choice for all babies. In infants with at least one first degree relative with atopy, hydrolyzed formula for a minimum of four months combined with dietary restrictions and environment measures may reduce the risk of developing asthma or wheeze in the first year of life. There is insufficient evidence to suggest that soya-based milk formula has any benefit [21]

Another meta-analysis on soy formula for prevention of allergy and food intolerance in infants found no significant difference in childhood allergy incidence (2 studies; typical relative risk: 0.73; 95% confidence interval: 0.37 to 1.44) [22]. No significant difference was reported in one study in infant asthma (relative risk: 1.10; 95% confidence interval: 0.86 to 1.40), infant eczema (relative risk: 1.20; 95% confidence interval: 0.95 to 1.52), childhood eczema prevalence (relative risk: 1.10; 95% confidence interval: 0.73 to 1.68), infant rhinitis (relative risk: 0.94; 95% confidence interval: 0.76 to 1.16) or childhood rhinitis prevalence (relative risk: 1.20; 95% confidence interval: 0.73 to 2.00). Meta-analysis found no significant difference in childhood asthma incidence (3 studies, 728 infants; typical relative risk: 0.71; 95% confidence interval: 0.26 to 1.92), childhood eczema incidence (2 studies, 283 infants; typical relative risk: 1.57; 95% confidence interval: 0.90 to 2.75) or childhood rhinitis incidence (2 studies, 283 infants; typical relative risk: 0.69; 95% confidence interval: 0.06 to 8.00). One study reported no significant difference in infant cow's milk protein intolerance (relative risk: 1.09; 95% confidence interval: 0.45 to 2.62), infant cow's milk allergy (relative risk: 1.09; 95% confidence interval: 0.24 to 4.86), childhood soy protein allergy incidence (relative risk: 3.26; 95% confidence interval: 0.36 to 29.17) and urticaria. No study compared soy formula to hydrolysed protein formula. The authors conclude that feeding with a soy formula cannot be recommended for prevention of allergy or food intolerance in infants at high risk of allergy or food intolerance.

Two trials evaluating formulas containing hydrolyzed protein for prevention of allergy and food intolerance in infants compared early, short-term hydrolyzed formula to human milk feeding [23]. No significant difference in infant allergy or childhood cow's milk allergy was reported. No eligible trial compared prolonged hydrolyzed formula to human milk feeding. Two trials compared early, short-term hydrolyzed formula to cow's milk formula feeding. No significant benefits were reported. One large quasi-random study reported a reduction in infant cow's milk allergy of borderline significance in low risk infants (relative risk: 0.62; 95% confidence interval: 0.38 to 1.00) [23]. Ten eligible studies compared prolonged feeding with hydrolyzed formula versus cow's milk formula in high risk infants. Meta-analysis found a significant reduction in infant allergy (seven studies, 2514 infants; typical relative risk: 0.79, 95% confidence interval: 0.66 to 0.94), but not in the incidence of childhood allergy (two studies, 950 infants, typical relative risk: 0.85; 95% confidence interval: 0.69 to 1.05). There was no significant difference in infant eczema (eight studies, 2558 infants; typical relative risk: 0.84; 95% confidence interval: 0.68 to 1.04), childhood eczema incidence (two studies, 950 infants, typical relative risk: 0.83: 95% confidence interval: 0.63 to 1.10), childhood eczema prevalence (one study, 872 infants; relative risk: 0.66: 95% confidence interval: 0.43 to 1.02), or infant or childhood asthma, rhinitis and food allergy. One study reported a significant reduction in infants with cow's milk allergy with confirmed atopy (relative risk: 0.36; 95% confidence interval: 0.15 to 0.89). Subgroup analysis of trials blinded to formula found no significant difference in infant allergy (four studies, 2156 infants; typical relative risk: 0.87; 95% confidence interval: 0.69, 1.08) or childhood allergy incidence (one study, 872 infants; relative risk: 0.91; 95% confidence interval: 0.73 to 1.14). No eligible trial examined the effect of prolonged hydrolyzed formula feeding on allergy beyond early childhood. There is evidence that preterm or low-birth-weight infants fed a hydrolyzed preterm formula have significantly reduced weight gain, but not in other growth parameters (head circumference or length). Studies in term infants report no adverse effects on growth. Subgroup analysis of trials of partially hydrolyzed versus cow's milk formula found a significant reduction in infant allergy (six studies, 1391 infants; typical relative risk: 0.79; 95% confidence interval: 0.65 to 0.97) but not childhood allergy, or infant or childhood asthma, eczema or rhinitis. Methodological concerns were the same as for the overall analysis. Analysis of trials of extensively hydrolyzed formula versus cow's milk formula found no significant differences in allergy or food intolerance. Infants fed extensively hydrolyzed formula compared with partially hydrolyzed formula had a significant reduction in food allergy (two studies, 341 infants; typical relative risk: 0.43: 95% confidence interval: 0.19 to 0.99), but there was no significant difference in all allergy or any other specific allergy incidence. Comparing extensively hydrolyzed casein containing formula with cow's milk formula, one study (431 infants) reported a significant reduction in childhood allergy incidence (relative risk: 0.72; 95% confidence interval: 0.53 to 0.97). Meta-analysis found a significant reduction in infant eczema (three studies, 1237 infants; typical relative risk: 0.71; 95% confidence interval: 0.51 to 0.97). One study reported a significant reduction in childhood eczema incidence (relative risk: 0.66: 95% confidence interval: 0.44 to 0.98) and prevalence (relative risk: 0.50: 95% confidence interval: 0.27 to 0.92). The authors conclude that there is no evidence to support feeding with a hydrolyzed formula for the prevention of allergy compared to exclusive breast feeding. In high risk infants who are unable to be completely breast fed, there is limited evidence that prolonged feeding with a hydrolyzed formula compared to a cow's milk formula reduces infant and childhood allergy and infant CMA. In view of methodological concerns and inconsistency of findings, further large, well designed trials comparing formulas containing partially hydrolyzed whey, or extensively hydrolyzed casein to cow's milk formulas are needed [23].

Food Allergy and Rhinoconjunctivitis

Research on food allergies and their associations with allergic rhinitis or allergic conjunctivitis is scarce. As aforementioned in the meta-analyses evaluating formulas containing soya or hydrolysed protein for prevention of allergy and food intolerance in infants, there are methodological concerns and inconsistency of findings [22,23]. Overall, no protective effects are demonstrated.

In a long-term retrospective analysis of related risk factors and association with concomitant allergic diseases, hen's egg sensitization was significantly related to the appearance of asthma and rhinoconjunctivitis [24]. The authors show that egg sensitization, severity of atopic dermatitis, and onset of rhinoconjunctiviis were positively related to the occurrence of asthma. Furthermore, their analysis shows that, although the appearance of rhinoconjunctivitis was proportional to the incidence of atopy and asthma, it was inversely related to the persistence of atopic dermatitis (corrected odds ratio confidence intervals <1) [24].

Rhinitis or rhinoconjunctivitis following inhalation of food dusts or vapor is not uncommon in patients with food allergy. Rhinitis/rhinocojunctivitis typically occurs in association with other clinical manifestations such as cutaneous and/or gastrointestinal symptoms during acute allergic reactions to foods. Rhinitis/rhinocojunctivitis as the sole manifestation of food allergy is quite uncommon.

Food Allergy and Childhood Eczema

The prevalence of atopic dermatitis has increased two to three folds over the past three decades in industrialized countries and there is evidence to suggest that this prevalence is continuing to increase [5,25-27]. The increase in prevalence may be due to increased access to medical care, improved recognition, better epidemiological reporting, or increased environmental allergens due to industrialization and pollution. Atopic dermatitis affects 10 to 20% of children and 1 to 3% of adults in the United States and Europe [28].

Food allergy plays an immunopathogenic role in 30 to 50% of children with moderate to severe atopic dermatitis [1,29-31]. Most children with food allergy react to only one or two of the most common allergens such as egg, cow milk, nut, peanut, soy, and wheat [1,3-5]. Measuring food-specific serum IgE antibodies to six foods (milk, egg, wheat, soy, peanut, fish) known to be the most allergenic in children, approximately one third of children with refractory, moderate-severe atopic dermatitis have IgE-mediated clinical reactivity to one or more of these food proteins. The prevalence of food allergy in this population is significantly higher than that in the general population, and an evaluation for food allergy should be considered in these patients [32]. For children whose food allergy has been identified, elimination of the offending allergen from the diet seems prudent. Conversely, avoidance of common foods in children without documented food allergy might result in faddism or malnutrition [33].

Results regarding the protective effect of breastfeeding vary widely [34,35]. The American Academy of Dermatology Guidelines Task Force reviewed the subject in 2004 and found no conclusive evidence that exclusive breastfeeding influences the development of atopic dermatitis [34]. There is, however, suggestive evidence that prolonged breastfeeding may delay the onset of atopic dermatitis [34]. In the same year, a group of experts of the Section of Pediatrics, the European Academy of Allergology and Clinical Immunology reviewed critically the existing literature and concluded that exclusive breastfeeding for at least 4 to 6 months in infants with atopic hereditary would result in a lower incidence of atopic dermatitis [35]. In high risk infants, exclusive breastfeeding for the first six months of life is recommended which may delay or prevent the onset of atopic dermatitis [35]. When breastfeeding is not possible, a partially or completely hydrolysed formula is desirable [36]. Allergen avoidance during pregnancy does not have a protective effect against developing atopic dermatitis and may lead to preterm births and reductions in birth weight. The present consensus is that dietary intervention in utero is potentially harmful and is not indicated [36].

In atopic eczema, food avoidance is extremely prevalent, especially by parents of infants and young toddlers. In a minority of infants with genuine allergy to cow's milk protein, the use of a hypoallergenic formula will ameliorate the severity of atopic dermatitis. Parents often try a soy formula. However, soy is another common allergen and the use of soy formula may not help. The disease often remains severe despite multiple changes of infant formulas and blind avoidance of many foods. In a small case series, a remarkable improvement in the severity of the atopic dermatitis was noted after sufficient explanation of the causes of atopic dermatitis and use of topical therapy [37]. The authors caution that it is a common misconception that food allergies and atopic dermatitis are always causally related. Alternatively, they point out that atopic dermatitis results from defective skin barrier function, for which topical treatment is essential, and the view is similarly shared by other authors [36]. Unjustified focus on food allergies as the primary cause of atopic dermatitis increases the risk of unnecessary dietary restriction with resultant malnutrition. Only children with acute allergic symptoms (for example, urticaria and gastrointestinal symptoms) directly related to food ingestion should be evaluated for food allergies by a double-blind, placebo-controlled food challenge.

The pattern of food avoidance is cultural-dependent. In a UK study, egg and diary products are often avoided by parents. In a Hong Kong study, parents often avoid giving seafood (especially crustaceans) and beef to their children with eczema. These foods are believed to be literally "poisonous" to children with eczema and various skin conditions. Many practitioners of alternative medicine ascribe symptoms to food allergy whereas other doctors do not. The causal relationships between some of these conditions and food allergies have not been evaluated extensively enough to provide sufficient evidence to become authoritative. Conversely, studies have demonstrated that the majority of children with allergies to milk, egg, soy, and wheat will outgrow their allergy.

Food allergy and atopic dermatitis in young children are believed to be more likely to be cell-mediated and have been evaluated using atopic patch tests [15,38,39]. Using more vigorous assessment with double-blind placebo-controlled oral food challenges in patients with more severe eczema, the role of food allergy may be more clear-cut. Food hypersensitivity is reported to play an immunopathogenic role in atopic dermatitis in approximately one-third of children. In 320 selected children with moderate to severe atopic dermatitis, 63% of children were found to have food hypersensitivity by double-blind placebo-controlled food challenges. Both IgE-mediated mast cell and mononuclear cell activation appear responsible for the eczematous lesions resulting from ingestion of food allergens [8]. Furthermore, 132 children with severe atopic dermatitis were evaluated for food hypersensitivity using double-blind placebo-controlled oral food challenges. Fifty-nine percent of the children experienced at least one immediate hypersensitivity response. Definitive diagnosis of food allergy and initiation of an appropriate elimination diet resulted in significant clinical improvement in the majority of patients with atopic dermatitis and food hypersensitivity [3].

In a long-term retrospective analysis of related risk factors and association with concomitant allergic diseases, atopic dermatitis had completely disappeared in 124 cases (60.5%). Other allergic manifestations that appeared included asthma in 70 cases (34.1%) and rhinoconjunctivitis in 118 cases (57.6%). Generally the average age of patients recovering from atopic dermatitis was higher in severe atopic dermatitis (6.0 ± 3.5 years) than in its moderate or mild forms (5.8 ± 4.5 and 5.5 ± 3.9 years, respectively). This phenomenon was particularly evident in children with hen's egg sensitization, who show a longer persistence of the condition (Student $t = 2.462$ and $P < .02$). The initial severity score of atopic dermatitis was found to be associated with a high frequency of asthma appearance (Pearson $\chi^2 = 14.225$ and $P < .001$). Hen's egg sensitization was significantly related to the appearance of asthma (Fisher's exact test $P < .007$) and rhinoconjunctivitis (Fisher's exact test $P < .05$). A retrospective analysis of relative risk factors and their association with concomitant allergic diseases in their case studies shows that the egg sensitization, severity of atopic dermatitis, and onset of rhinoconjunctivitis were positively related to the occurrence of asthma. In addition, their analysis shows that, although the appearance of rhinoconjunctivitis was proportional to the incidence of atopy and asthma, it was inversely related to the persistence of atopic dermatitis. The authors conclude that the use of defined criteria of clinical diagnosis for the determination of the condition's severity, along with the performance of objective allergometric tests at the time of inclusion, shows that the course of atopic dermatitis is significantly related to egg sensitivity. In addition, the average healing time is higher in egg-sensitive patients affected by the most severe form of atopic dermatitis than in mild or moderate cases [24].

Conclusions

Food allergy may be associated with eczema and occasionally with airway atopies. It should be accurately diagnosed and evaluated by a combination of tests with or without food challenge in young children with more severe disease. In the less severe cases, food sensitization/allergy may not be associated with disease severity and vigorous investigations are probably not indicated. It is important to evaluate whether the atopic patients are genuinely "allergic" to some of these food items. Management is suboptimal if children with food allergy and severe disease continue to consume the culprit food. Conversely, avoidance of common foods in children without food allergy could result in food faddism or malnutrition.

References

[1] Sampson, H A. Food allergy - accurately identifying clinical reactivity. *Allergy*. 2005; 60 (Suppl 79):19-24.

[2] Sicherer, SH; Sampson, HA. 9. Food allergy. *J. Allergy Clin. Immunol.* 2006; 117(Suppl 2):S470-S475.

[3] Sampson, HA. Food hypersensitivity as a pathogenic factor in atopic dermatitis. *N. Engl. Regional Allergy Proc*. 1986; 7:511-519.

[4] Hon, KL; Leung, TF; Lam, MC; et al. Eczema exacerbation and food atopy beyond infancy: how should we advise Chinese parents about dietary history, eczema severity, and skin prick testing? *Adv. Therapy* 2007; 24:223-230.

[5] Leung, AK; Hon, KL; Robson, WL. Atopic dermatitis. *Adv. Pediatr* 2007; 54:241-273.

[6] Venter, C; Pereira, B; Grundy, J; et al. Incidence of parentally reported and clinically diagnosed food hypersensitivity in the first year of life. *J. Allergy Clin. Immunol.* 2006; 117:1118-1124.

[7] Isolauri, E; Turjanmaa, K. Combined skin prick and patch testing enhances identification of food allergy in infants with atopic dermatitis. *J. Allergy Clin. Immunol*. 1996; 97:9-15.

[8] Sampson, HA. The immunopathogenic role of food hypersensitivity in atopic dermatitis. *Acta Derm.Venereol.*1992; 176:534-537.

[9] Rook, GA; Stanford, JL. Give us this day our daily germs. *Immunol. Today* 1998; 19:113-116.

[10] Rook, GA; Brunet, LR. Give us this day our daily germs. *Biologist*. 2002; 49:145-149.

[11] Liu, AH. Hygiene theory and allergy and asthma prevention. *Paediatr. Perinat. Epidemiol.* 2007; 21 (Suppl 3):2-7.

[12] Williams, H; Flohr, C. How epidemiology has challenged 3 prevailing concepts about atopic dermatitis. *J. Allergy Clin. Immunol.* 2006; 118:209-213.

[13] Verstege, A; Mehl, A; Rolinck-Werninghaus, C; et al. The predictive value of the skin prick test wheal size for the outcome of oral food challenges. *Clin. Exp. Allergy.* 2005; 35:1220-1226.

[14] Hill, DJ; Heine, RG; Hosking, CS. The diagnostic value of skin prick testing in children with food allergy. *Pediatr. Allergy Immunol.* 2004; 15:435-441.

[15] Roehr, CC; Reibel, S; Ziegert, M; et al. Atopy patch tests, together with determination of specific IgE levels, reduce the need for oral food challenges in children with atopic dermatitis. *J. Allergy Clin. Immunol.* 2001;107:548-553.

[16] Mehl, A; Rolinck-Werninghaus, C; Staden, U; et al. The atopy patch test in the diagnostic workup of suspected food-related symptoms in children. *J. Allergy Clin. Immunol.*2006; 118:923-929.

[17] Stromberg, L. Diagnostic accuracy of the atopy patch test and the skin-prick test for the diagnosis of food allergy in young children with atopic eczema/dermatitis syndrome. *Acta Paediatr.* 2002; 91:1044-1049.

[18] Heine, RG; Verstege, A; Mehl, A; et al. Proposal for a standardized interpretation of the atopy patch test in children with atopic dermatitis and suspected food allergy. *Pediatr. Allergy Immunol.* 2006; 17:213-217.

[19] Rhodes, HL; Sporik, R; Thomas, P; et al. Early life risk factors for adult asthma: a birth cohort study of subjects at risk. *J. Allergy Clin. Immunol.* 2001; 108:720-725.

[20] Berns, SH; Halm, EA; Sampson, HA; et al. Food allergy as a risk factor for asthma morbidity in adults. *J. Asthma.* 2007; 44:377-381.

[21] Ram, FS; Ducharme, FM; Scarlett, J. Cow's milk protein avoidance and development of childhood wheeze in children with a family history of atopy. *Cochrane Database of Sys. Rev.* 2007;(2):CD003795.

[22] Osborn, DA; Sinn, J. Soy formula for prevention of allergy and food intolerance in infants. *Cochrane Database Syst. Rev.* 2006;(4):CD003741.

[23] Osborn, DA; Sinn, J. Formulas containing hydrolysed protein for prevention of allergy and food intolerance in infants. Cochrane Database Syst. Rev. 2006;(4):CD003664.

[24] Ricci, G; Patrizi, A; Baldi, E; et al. Long-term follow-up of atopic dermatitis: retrospective analysis of related risk factors and association with concomitant allergic diseases. *J. Am. Acad. Dermatol.* 2006; 55:765-771.

[25] Williams, HC. Is the prevalence of atopic dermatitis increasing? *Clin. Exp. Dermatol.* 1992; 17:385-391.

[26] Leung, DY; Boguniewicz, M; Howell, MD; et al. New insights into atopic dermatitis. *J. Clin. Investig.* 2004; 113:651-657.

[27] Grillo, M; Gassner, L; Marshman, G; et al. Pediatric atopic eczema: the impact of an educational intervention. *Pediatr. Dermatol.* 2006; 23:428-436.

[28] Leung, DY; Nicklas, RA; Li, JT; et al. Disease management of atopic dermatitis: an updated practice parameter. Joint Task Force on Practice Parameters. *Ann. Allergy Asthma Immunol.* 2004; 93(3 Suppl 2):S1-21.

[29] Leung, AK; Barber, KA. Managing childhood atopic dermatitis. *Adv. Ther.* 2003; 20:129-137.

[30] Levy, RM; Gelfand, JM; Yan, AC. The epidemiology of atopic dermatitis. *Clin. Dermatol.* 2003; 21:109-115.

[31] Sampson, HA. Food sensitivity and the pathogenesis of atopic dermatitis. *J. R. Soc. Med.* 1997; 90 (Suppl 30):2-8.

[32] Eigenmann, PA; Sicherer, SH; Borkowski, TA; et al. Prevalence of IgE-mediated food allergy among children with atopic dermatitis. *Pediatrics* 1998; 101(3):E8.

[33] Hon, KL; Leung, TF; Kam, WY; et al. Dietary restriction and supplementation in children with atopic eczema. *Clin. Exp. Dermatol.* 2006; 31:187-191.

[34] Hanifin, JM; Cooper KD; Ho, VC; et al. Guidelines of care for atopic dermatitis, developed in accordance with the American Academy of Dermatology (AAD)/American Academy of Dermatology Association "Administrative Regulations for Evidence-Based Clinical Practice Guidelines". *J. Am. Acad. Dermatol.* 2004; 50:391-404.

[35] Muraro, A; Dreborg, S; Halken, S; et al. Dietary prevention of allergic diseases in infants and small children. Part III: Critical review of published peer-reviewed observational and interventional studies and final recommendations. *Pediatr. Allergy Immunol.* 2004; 15:291-307.

[36] Simpson, EL. Atopic dermatitis prevention. *Dermatol. Ther.* 2006; 19:108-117.

[37] Wensink, M; Timmer, C; Brand, PL. [Atopic dermatitis in infants not caused by food allergy]. [Dutch]. *Nederlands Tijdschrift voor Geneeskunde* 2008; 152:4-9.

[38] Kalach, N; Soulaines, P; de Boissieu, D; et al. A pilot study of the usefulness and safety of a ready-to-use atopy patch test (Diallertest) versus a comparator (Finn Chamber) during cow's milk allergy in children. *J. Allergy Clin. Immunol.* 2005; 116:1321-1326.

[39] Keskin, O; Tuncer, A; Adalioglu, G; et al. Evaluation of the utility of atopy patch testing, skin prick testing, and total and specific IgE assays in the diagnosis of cow's milk allergy. *Ann. Allergy Asthma Immunol.* 2005; 94:553-560.

[40] Hill, DJ; Hosking, CS. Food allergy and atopic dermatitis in infancy: an epidemiologic study. *Pediatr. Allergy Immunol.* 2004; 15:421-427.

[41] Giusti, F; Seidenari, S. Patch testing with egg represents a useful integration to diagnosis of egg allergy in children with atopic dermatitis. *Pediatr. Dermatol.* 2005; 22:109-111.

[42] Niggemann, B. The atopy patch test (APT) - a useful tool for the diagnosis of food allergy in children with atopic dermatitis. *Allergy*. 2000; 55:281-285.

[43] Seidenari, S; Giusti, F; Bertoni, L; et al. Combined skin prick and patch testing enhances identification of peanut-allergic patients with atopic dermatitis. *Allergy*. 2003; 58:495-499.

In: Food Allergies: New Research
Editor: Carrie M. Chesterton, pp. 135-156

ISBN 978-1-60456-978-0

Chapter 4

MANAGEMENT OF THE CHILD WITH FOOD ALLERGY

***Alexander K.C. Leung*[1,*] *and Kam-lun Ellis Hon*[2]**
[1] University of Calgary, Alberta Children's Hospital,
Calgary, Alberta, T2M 0H5, Canada
[2] Chinese University of Hong Kong, Prince of Wales Hospital,
Shatin, Hong Kong

ABSTRACT

The definitive treatment of food allergy is strict elimination of the offending food from the diet. Symptomatic reactivity to food allergens is generally very specific, and patients rarely react to more than one food in a botanical or animal species. If elimination diets are prescribed, care must be taken to ensure that they are palatable and nutritionally adequate. Patients must have a good knowledge of food containing the allergen and must be taught to scrutinize the labels of all packaged food carefully.

Formula-fed infants with cow's milk allergy should be fed an elemental or extensively hydrolysed hypoallergenic formula. Soy formulas are inappropriate alternatives as a significant number of infants who are allergic to cow's milk are also allergic to soy. Most children outgrow their food hypersensitivity. As such, rechallenge testing for food allergy should be performed; the interval between rechallenges should be dictated by the specific food allergen in question, the age of the child, and the degree of difficulty in avoiding the food in question

Emergency treatment of food-induced anaphylaxis should follow the basic life support ABC principles, with the simultaneous intramuscular injection of adrenaline. A fast-acting H_1 antihistamine should be considered for the child with progressive or generalized urticaria or disturbing pruritus. Pharmacological therapies such as mast cell stabilizers have very little role to play in the treatment of gastrointestinal manifestations of food allergy.

In high-risk infants, exclusive breastfeeding with introduction of solid foods not earlier than 6 months of age may delay or possibly prevent the onset of food allergy in some children. Avoidance of allergenic foods by lactating mothers is often recommended. When breastfeeding is not possible, the use of a partially or extensively

---

* Correspondence to: Dr. Alexander K.C. Leung, #200, 233 – 16th Avenue NW, Calgary, Alberta, Canada T2M 0H5; Telefax: (403) 230-3322, e-mail: aleung@ucalgary.ca

hydrolysed hypoallergenic formula is desirable. Prophylactic medications have not been shown to be consistently effective in the prevention of life-threatening reactions to food. Their use may mask a less severe reaction to a culprit food, knowledge of which might prevent a more severe reaction to that food in the future.

Keywords: allergen avoidance, breastfeeding, hydrolysed hypoallergenic formula

Introduction

Food allergy affects approximately 4 to 8% of young children and 1 to 4% of adults [1-4]. The key to the management of food allergy is avoidance of foods known to or suspected of having caused a reaction [5]. Other approaches include pharmacotherapy, education of patients, and dietary manipulations for the prevention of food allergy in high-risk individuals. Currently, there is no effective and safe immunotherapy in the management of patients with food allergy.

Avoidance of Food Allergens

The definitive treatment of food allergy is strict elimination of the offending food from the diet [6,7]. It is unusual for a child to be allergic to more than one food. Bock studied 480 children with probable food hypersensitivity and found that allergic reaction to more than two foods occurred in only 10 (2.1%) of the 480 children [8]. Crespo et al. studied 355 children with food allergy and found that 222 (62.5%) had allergic reactions to only one food, 86 (24.2%) had allergic reactions to two foods, and 47 (13.2%) had allergy to three or more foods [9]. Symptomatic reactivity to food allergens is generally very specific, and patients rarely react to more than one food in a botanical or animal species [7,10-12]. However, in pollen-related food allergy, cross-reactions can occur between phylogenetically distantly related species such as birch and kiwi or soy [2].

The avoidance of the offending food antigen sounds simple but in reality it is not, especially if the offending food is ubiquitous and thus difficult to avoid. Avoidance can be difficult because of cross-contamination of food which may lead to inadvertent ingestion of the offending food [13]. Accidents most frequently happen in daycare centers, schools, and restaurants [13,14]. Approximately 50% of affected individuals experience accidental exposure and reactions every 3 to 5 years [15]. Of the 32 food-related fatalities reported by Bock et al, at least 87% of patients had a previous history of reaction to the responsible food allergen [16]. Avoidance of skin contact and inhalation of offending food allergen is also necessary [17-19]. The importance of education about allergen avoidance cannot be over-emphasized.

On the other hand, the indiscriminate use of elimination diets without a firm diagnosis is a widespread malpractice and may lead to psychological dependence on an unsound diet, as well as vitamin deficiencies, malnutrition, and failure to thrive if multiple foods are inadvertently avoided [7,20-23]. The diagnosis of food allergens should be based on placebo-controlled food challenges. Venter et al followed a birth cohort of 966 infants on the Island of Wight, United Kingdom born between September 2001 and August 2002 to the age of one year [24]. Cumulative incidence of parentally reported food hypersensitivity was 25.8%. Open or double-blind, placebo-controlled food challenges were used to confirm suspected reactions. Only 2.2% of those tested were confirmed to have food allergy, indicating the need to evaluate suspected food allergy to avoid needless dietary restriction.

Oral allergy syndrome generally occurs in patients with inhalant allergy to birch, mugwort, or ragweed pollen and is associated with the ingestion of various fresh fruits (e.g., bananas, melons, citrus fruits) and raw vegetables (e.g. carrots, tomatoes, celery) [13,25]. It is interesting to note that if the offending fruit or vegetable is cooked, then the patient does not usually experience any symptom as the food allergens are generally destroyed by heating [26]. Thus, avoiding the offending fruit or vegetable which has been cooked may not always be necessary.

In some children ingesting ludicrous elimination diets, eating disorders may develop. If elimination diets are prescribed, care must be taken to ensure that they are palatable and nutritionally adequate. Patients should be provided with information on what alternative foods are available so that good variety in the diet can be maintained [27]. A formal dietetic evaluation is recommended.

Patients and/or their caregivers must have a good knowledge of foods containing the allergen and must be taught to scrutinize the labels of all packaged food carefully [28]. Careful label reading is the cornerstone of food allergy management [29]. In one study, only 4 (7%) of 60 parents were able to identify milk protein in 14 sample labels [30]. Incorrect or ambiguous labeling of foods may result in accidental ingestion of the offending food [13]. Also, some of the terms used do not clearly indicate the presence of a food allergen. The United States Food and Drug Administration (FDA) now requires food manufacturers to declare and clear label all functional ingredients on food labels [31]. The Food Allergen Labeling and Consumer Protection Act (FALCPA) effective in January 2006 requires simple terms to indicate the presence of major food allergens [4]. FALCPA requires food manufacturers to state explicitly the presence of eight major food allergens, namely, milk, egg, wheat, soybean, peanut, tree nuts, fish, and shellfish. New EU labeling laws require the presence of the following food allergens at any level to be stated on the label: celery, cereals containing gluten (wheat, barley, rye and oats), crustaceans, eggs, fish, milk, mustard, tree nuts, peanuts, sesame seeds, soybeans, SO_2 and sulfites (at level >10 mg/kg or >10 mg/L) [32]. It is mandatory to list all subgredients and specify the source of ingredients previously listed as "natural flavor". However, foods that are not prepacked are not covered by this legislation [27]. Patients and/or their caregivers should be cautioned about the presence of the offending food as a "hidden" ingredient in processed foods [33,34]. The importance of communicating with restaurant staff about ingredients when one is dining out cannot be over-emphasized [35].

Cow's milk allergy affects 0.3 to 7.5% of infants [36]. The incidence of self-diagnosed cow's milk allergy is substantially higher than that reported in randomized, controlled, food challenge trials [37]. Infants with cow's milk protein allergy should not be fed either whole cow's milk or formulas containing intact whole cow's milk proteins [36]. It has been estimated that 15 to 25% of infants who have IgE-mediated cow's milk allergy are also allergic to soy, but the rate of tolerance is only 50% for those with non-IgE-mediated cow's milk allergy [4,38,39]. A Cochrane analysis of studies comparing soy to hydrolysed cow's milk formulas found a significant increase in infant and childhood allergy cumulative incidence in infants fed soy formulas [40]. As such, soy formula is not a suitable alternative [37]. Goat's milk is not recommended in infants with cow's milk allergy as goat's milk also shares some allergenic protein fractions with cow's milk [37,41,42]. Infants with cow's milk or soy hypersensitivity should be fed a hypoallergenic formula [43]. Extensively hydrolysed casein formulas such as Nutramigen, Pregestimil, and Alimentum have also been used successfully in this regard [43,44]. These formulas are hypoallergenic and well tolerated by children [45]. These formulas, however, are expensive and unpalatable. The partially hydrolyzed whey hydrolysate Good Start is less expensive and has a better taste [46]. However, it contains slightly larger peptides and significantly more immunologically identifiable cow's milk protein, and therefore not suitable for the treatment of cow's milk allergy [1,39,47]. Formulas whose protein source is free amino acids (e.g., Vivonex, Ele Care, and Neocate) are available and are considered as nonallergenic [1,42]. These formulas should be tried in infants who are very sensitive to cow's milk protein and cannot tolerate even extensively hydrolysed formulas [14]. Amino acid-based formulas are also useful in the treatment of allergic eosinophilic esophagitis and allergic eosinophilic gastroenteropathy [48-51].

Breastfed infants who develop symptoms of food allergy may benefit from maternal restriction of cow's milk, egg, fish, peanuts, and tree nuts [52]. Approximately 50% of infants who have food protein-induced proctocolitis while nursing improve with elimination of cow's milk from the maternal diet [53]. If maternal restriction of food allergens is not successful, an extensively hydrolysed formula should be tried [52]. If allergic symptoms persist, a free amino acid-based formula should be considered [52].

Most children outgrow their food hypersensitivity [46]. In contrast, the prognosis for adults with food allergy is less favorable [2]. It has been shown that approximately 30 to 40% of children lose their food hypersensitivity after 1 to 2 years of allergen avoidance, even though the results of skin tests and radioallergosorbent tests may not change [54]. The loss of hypersensitivity is especially likely to occur in infants and young children, although older children and adults may also lose their hypersensitivity [55]. The degree of compliance with allergen avoidance and the allergen responsible may influence the outcome [46]. The majority of children outgrow cow's milk allergy by 4 years of age [42]. Hypersensitivity to peanuts, nuts, egg, fish, and shellfish tends to be more persistent [27,42,54,56]. Consequently, rechallenge testing for food allergy should be performed; the interval between rechallenges should be dictated by the specific food allergen in question, the age of the child, and the degree of difficulty in avoiding the food in question [12,47].

PHARMACOTHERAPY

Symptomatic treatment of complications resulting from the inadvertent ingestion of food is essentially the same as that for the specific complication resulting from any other cause [57]. Patients with a history of anaphylactic reaction over the age of seven years as well as caregivers should be taught how to self-administer epinephrine and should have an epinephrine auto-injector such as Anapen/Anapen Jr, EpiPen/EpiPen Jr, or Twinjet/Twinjet Jr and antihistamine available at all times [2,58]. The physician should take appropriate steps to ensure that the patients and their caregivers understand the indications and use of the device thoroughly. These individuals should also be provided with a written anaphylaxis action plan. The instructions should be clear, simple, and age appropriate. Rehearsal of the procedure is important.

Epinephrine helps to block severe allergic reactions and anaphylaxis by suppressing leukotriene and histamine release [59,60]. Epinephrine reverses vasodilatation, increases blood pressure, dilates airways, reduces laryngeal edema and angioedema, and increases myocardial contractility [59,60]. For the treatment of anaphylaxis, the recommended dose of epinephrine 1:1000 (1mg/ml) is 0.01 mg/kg intramuscularly, up to a maximum of 0.3 mg (0.3 ml) in children and 0.5 mg (0.5 ml) in adults [14,59,61]. Peak concentrations are reached within 10 minutes of intramuscular administration [60,62]. As Anapen Jr, EpiPen Jr, and Twinjet Jr contain 0.15 mg of epinephrine and Anapen, EpiPen, and Twinjet contain 0.3 mg of epinephrine, it would be desirable to have a wider range of auto-injector doses [13,59]. The subcutaneous route is no longer recommended as the systemic levels of epinephrine are highly unpredictable from this mode of administration [59]. It has been shown that intramuscular injections into the thigh result in more rapid absorption and higher plasma epinephrine levels than intramuscular injections into the arm [28,61,63-65]. Epinephrine works best when given early [13,59].

After the first aid treatment, the patient should be transferred to the nearest emergency department for monitoring and additional treatment as required [14]. If necessary, the dose of epinephrine may be repeated at 5 minutes intervals according to cardiorespiratory function [59,64]. In one study, 16% of patients presenting to the emergency department with food-induced anaphylaxis required two doses of epinephrine [64]. In patients with severe anaphylaxis unresponsive to intramuscular epinephrine or with cardiovascular collapse, epinephrine should be given intravenously with blood pressure and continuous cardiac monitoring [60,61]. Some of these patients may require volume support, oxygen, nebulized bronchodilators, parenteral diphenhydramine, ranitidine, and glucocorticosteroids [28,66]. Should this be the case, the patient should be placed in a recumbent position with lower limbs elevated, as tolerated symptomatically [28,61,63]. This may prevent orthostatic hypotension. Because of the possibility of a biphasic response, all patients with anaphylaxis should be observed for at least four hours before discharge.

Patients and/or their caregivers must be educated about early recognition of allergic symptoms and early management of an anaphylactic reaction [67]. Schools should be equipped to treat anaphylaxis in allergic students and physicians should help instruct school personnel about these issues [2]. The Food Allergy & Anaphylaxis Network (FAAN) (10,400 Easton Place, Fairfax, VA 22030-5647; tel.800-929-4040; or www.foodallergy.org) has excellent educational material for patients, schools, and physicians dealing with food allergy in addition to a written emergency plan for treatment of an accidental ingestion [12]. It is suggested that patients at risk for anaphylaxis should always carry two doses of self-injectable epinephrine [64,68]. An identification necklace or bracelet such as MedicAlert or Medi-Tag stating the patient's sensitivity is also advised [1,34]. Referral to an allergist is also recommended [69].

For the child with progressive or generalized urticaria or distressing pruritus, the administration of a fast-acting oral H_1 antihistamine such as hydroxyzine or diphenhydramine should be considered [10,59,61,70]. H_2 receptor blockers such as ranitidine are less helpful as only a small number of H_2 receptors are found in the skin [70]. Although antihistamines do not block systemic reactions, they do relieve the itchiness [67].

The use of drugs such as disodium cromoglycate, ketotifen, and prostaglandin synthetase inhibitors in the treatment of food allergy has generally been disappointing, either because of minimal efficacy or unacceptable adverse effects [7,12]. Systemic corticosteroids are rarely used in the treatment of food allergy, except in severe anaphylaxis, allergic eosinophilic esophagitis, allergic eosinophilic gastroenteropathy, and dietary-induced enteropathy [67,71]. The side effects of long-term systemic corticosteroid therapy are unacceptable.

Currently, oral administration of activated charcoal is not considered a practical first-aid treatment for food anaphylaxis [14]. Prophylactic medications have not been shown to be consistently effective in the prevention of severe life-threatening reactions to foods [5]. Their use may mask a less severe allergic reaction to a culprit food, the knowledge of which might prevent a more severe allergic reaction to that food in the future [5]. The use of prophylactic medications is therefore discouraged.

PROBIOTICS AND PREBIOTICS

The use of probiotics and prebiotics in the management of food allergy is controversial [72]. It has been hypothesized that the increased sensitization to food allergens might result from reduced infection or exposure to microbial products such as endotoxin in early childhood [73]. Prospective studies have found that infants who are prone to develop atopic dermatitis have lower numbers of *Bifidobacterium* in their intestinal microflora [74-76]. It has been shown that probiotics might reverse the increased intestinal permeability characteristic of children with food allergy and enhance specific IgA responses frequently defective in children with food allergy [77]. *In vitro* studies show that allergic patients induce less IL-10 production and more proinflammatory cytokine production than those nonallergic individuals [78,79]. Presumably, probiotics act on the intestinal mucosa and stimulate T-cell differentiation in favor of Th1 over Th2, with resultant decreased production of IgE and increased production of IgA [1,78,80]. Probiotics might also correct aberrations in gut permeability [81,82]. Prebiotics work by selectively stimulating the growth or activity of a limited number of bacterial strains in the intestinal flora.

Much work has been done on the use of probiotics and, to a lesser extent, prebiotics in the management of atopic dermatitis. Several randomized controlled trials failed to show the beneficial effects of probiotics in the prevention of atopic dermatitis [83]. Other studies yielded different results [83-88]. Kalliomäki et al. randomized 159 mothers and their respective infants with a family history of atopy to receive either a placebo or 10^{10} CFU of *Lactoobacillus* GG for 2 to 4 weeks before delivery and for 6 months after delivery, respectively [84]. Twenty three percent of the children in the probiotic group versus 46% of children in the control group were found to have atopic dermatitis at two years of age (RR: 0.51; 95% confidence interval: 0.32-0.84) [84]. The effect was still observed two years later: 26% of children in the treatment group versus 46% of children in the placebo group had atopic dermatitis (RR: 0.57; 95% confidence interval: 0.33-0.97) [85].

Viljanen et al. randomized 230 infants who had suspected cow's milk allergy in a double-blinded study to receive *L. rhamnosus* GG (n=80), a mixture of four probiotic strains (n=76), or a placebo (n=74), given twice daily with food for four weeks [88]. The authors found that

L. rhamnosus GG was an effective therapy for atopic dermatitis in IgE-sensitized infants but not in non-IgE-sensitized infants.

In a double-blind, placebo-controlled trial, Tamura et al. randomized 109 adult patients with allergic rhinitis to drink fermented milk containing *Lactobacilli casei* strain Shirota (n=55) or placebo (n=54) for 8 weeks [89]. The authors found no significant difference between the two groups during the ingestion period. In the subgroup of patients with moderate to severe nasal symptom scores before starting ingestion of test samples, supplementation with the probiotic tended to reduce nasal symptom-medication scores.

In a double-blind placebo-controlled trial, Taylor et al randomized 226 newborn infants of atopic mothers to receive either 3 x 10^9 CFU of *Lactobacillus acidophilus* (n=115) or placebo (n=111) daily for 6 months [90]. A total of 178 infants (89 in each group) completed the study. The authors found that the rates of atopic dermatitis were similar in the two groups at 6 months and 12 months of follow-up. At 12 months, the rate of sensitization was significantly higher in the probiotic group (p=0.03). These findings challenge the use of probiotics in the prevention of allergy.

Kukkonen et al. randomized 1,223 pregnant women carrying high risk infants at increased risk for allergy to receive a probiotic (n=610) or a placebo (n=613) for 2 to 4 weeks before delivery [91]. Their infants received the same probiotic plus galacto-oligosaccharides (n=461) or a placebo (n=464) for 6 months. These children were evaluated at 2 years of age for cumulative incidence of allergic diseases (food allergy, eczema, asthma, and allergic rhinitis) and IgE sensitization (positive skin prick test response or serum antigen-specific IgE level). The authors found that probiotic and prebiotic treatment, compared with placebo, had no effect on the cumulative incidence of allergic diseases but tended to reduce IgE-associated atopic diseases (odds ratio: 0.71; 95% confidence interval: 0.5 to 1; p=0.052). Probiotic and prebiotic treatment did reduce eczema (odds ratio: 0.74; 95% confidence interval: 0.55 to 0.98; p=0.35) and atopic eczema (odds ratio: 0.66; 95% confidence interval: 0.46 to 0.95; p=0.025).

In a double-blind placebo-controlled trial, Weston et al. randomized 56 children aged 6 to 18 months who had moderate to severe atopic dermatitis to receive *L. fementum* VRI-033 PCC (n=28) or placebo (n=28) twice daily for eight weeks [92]. Fifty children completed the study. The authors found that the reduction in the SCORAD index was significant in the probiotic group (p=0.03) but not in the placebo group. In a double-blind study, Passeron et al. randomized 48 children to receive either *L. rhamnosus* Lcr 35 plus a prebiotic preparation (n=28) or an identically appearing probiotic preparation alone three times a day for three months [93]. In the symbiotic group, the mean total SCORAD score was 39.1 before treatment versus 20.7 after three months of treatment (p< 0.0001). In the probiotic group, the mean SCORAD score was 39.3 before treatment versus 24 after three months of treatment (p<0.0001). The authors concluded that symbiotics and prebiotics used alone were effective in the treatment of atopic dermatitis.

Probiotics and prebiotics are included in some infant formulas with the aim of inducing the development of a *Bifidobacterium*-dominated intestinal flora [37]. At present, probiotics or prebiotics are not established treatment modalities for atopic dermatitis [72,78]. They are ineffective in the prevention and treatment of reactive airway disease [78]. The routine use of probiotics and prebiotics in food allergy management requires further study.

Dietary Manipulations in the Prevention of Food Allergy in High-Risk Individuals

Breastfeeding has a protective effect on the incidence of atopic disease in children who have a genetic predisposition for atopy [53,94-97], especially so for the prevention of asthma [95]. Gdalevich et al. performed a systematic review and meta-analysis of 12 prospective studies (n=8,183) that evaluated the association between exclusive breastfeeding in the first three months of life and asthma [95]. The mean follow-up was 4.1 years. The pooled odds ratio for the protective effect of breastfeeding was 0.7 (95% confidence interval: 0.6 to 0.81). The protective effect was more pronounced in the subset of studies that assessed children with a family history of atopy (odds ratio: 0.52; 95% confidence interval: 0.35 to 0.79) compared to studies with an unstratified population, and to studies that included children without atopic first-degree relatives (odds ratio: 0.73; 95% confidence interval: 0.62 to 0.86).

The protective effects of breastfeeding on atopic dermatitis are controversial. The American Academy of Dermatology Guidelines Task Force reviewed the subject in 2004 and found no conclusive evidence that exclusive breastfeeding influences the development of atopic dermatitis [98]. The American Academy of Dermatology Guidelines Task Force did find suggestive evidence that prolonged breastfeeding might delay the onset of atopic dermatitis [98]. The Section of Pediatrics of the European Academy of Allergology and Clinical Immunology critically reviewed the existing literature and concluded that exclusive breastfeeding for at least 4 to 6 months in infants with a family history of atopy results in a lower incidence of atopic dermatitis [99]. A recent study reported that breastfeeding for 4 months or more reduces the risk of atopic dermatitis and the onset of the allergy to 4 years of age [100]. Kull et al followed a birth cohort of 4,089 children [100]. Data on breastfeeding, allergic symptoms, and potential confounding factors were obtained from questionnaires that were answered by parents when the children were 2 months, and 1, 2, and 4 years of age. The authors found that exclusive breastfeeding for 4 months or more reduced the risk of atopic dermatitis at 4 years of age (odds ratio: 0.78; 95% confidence interval: 0.63 to 0.96). The decreased risk was most evident for children who had atopic dermatitis with an onset during the first 2 years of life and that persisted to 4 years (odds ratio: 0.59; 95% confidence interval: 0.45 to 0.77). A protective effect of breastfeeding was also noted among children with early onset atopic dermatitis, regardless of whether it was persistent, and among children who developed early or late onset asthma which was followed by late onset atopic dermatitis (odds ratio: 0.48; 95% confidence interval: 0.3 to 0.76).

Breastfeeding might have a protective effect on the development of allergic rhinitis, although not to the same extent as that reported for asthma and atopic dermatitis [95,100,101]. Minouni Bloch et al. performed a meta-analysis of six prospective studies (n=3,303) that examined the effect of exclusive breastfeeding in the first 3 months of life on the development of allergic rhinitis [101]. The overall odds ratio was 0.74 (95% confidence interval: 0.54 to 1.01). The protective association was considered substantial notwithstanding borderline statistical significance. More recently, Siltanen et al. prospectively followed 456 Finnish children from an unselected birth cohort of 4,674 infants for 4 years [102]. The authors found in children with a family history of atopy, exclusive breastfeeding for 3 months protected against allergic rhinoconjunctivitis and sensitization to pets [102].

In high-risk infants who are exclusively breastfeed and in whom solids are not introduced until six months of age, breastfeeding might delay, or possibly prevent, the onset of food allergy [6,23,39,103,104]. Infants with elevated cord serum IgE and a positive family history of atopy are at risk for the development of atopic disease [105-107]. Breastfeeding protects against the development of allergy by several potential mechanisms. Colostrum provides a protective coating to the gut that prevents the entrance of large foreign proteins and minimizes the possibility of an allergic response. Colostrum also prevents the adherence of pathogens. Breastfeeding reduces the amount of foreign protein in the gastrointestinal tract and passively transfers maternal IgA to the infant, which minimizes the risk for absorption of antigens from the gastrointestinal tract [6,108]. Transfer of cell-mediated immunity from mother to infant stimulates IgA synthesis in the infant [109]. Epidermal growth factor present in human milk hastens maturation of intestinal mucosa and epithelium, and strengthens the mucosal barrier to antigen [108]. Several studies have shown that respiratory and gastrointestinal infections can predispose to the development of allergic diseases [109]. The allergy-preventive effect of breastfeeding might be secondary to a reduction in the number of infections in the infant.

Because small amounts of food antigens ingested by the mother are excreted in breast milk [55], avoidance of allergenic foods by lactating mothers is often recommended [6,33,36,110,111]. A meta-analysis of two studies [112,113] concludes that avoidance of allergenic foods by lactating mothers may transiently reduce the development of atopic dermatitis in early childhood [111]. Several studies suggest that a strict food allergen avoidance regimen followed by mothers during pregnancy may result in a significant reduction in the infants of food-related allergic disorders [06,114,115]. Other authors disagree [105,116,117]. Studies that have demonstrated a beneficial effect of a strict food allergen avoidance regimen followed by mothers during pregnancy and lactation on the prevalence and severity of atopic disease in high-risk, breast-fed infants probably reflect only the influence of maternal dietary restriction during lactation [105]. The present consensus is that dietary intervention in utero is potentially harmful and is not indicated unless future studies prove otherwise [11,37,111,118].

Marini et al. prospectively studied 279 infants with high atopic risk who were put on an allergy prevention program and 80 infants with similar atopic risk but no intervention [119]. The intervention program included dietary measures (exclusive and prolonged milk feeding followed by a hypoallergenic weaning diet) and environmental measures such as avoidance of parental smoking in the presence of the babies. The incidence of allergic manifestations was much lower in the intervention group than in the nonintervention group at 1 year (11.5% vs. 54.4%), 2 years (14.9% vs. 65.6%), and 3 years (20.6% vs. 74.1%). Atopic dermatitis and recurrent wheezing were found in both the intervention group and the nonintervention group from birth to the second year of life, whereas urticaria and gastrointestinal disorders were only present in the nonintervention group in the first year of life. Halken et al. studied 105 "high-risk" infants who were breast-fed and/or receiving a hypoallergenic formula combined with avoidance of solid foods during the first 6 months of life [120]. This prevention group was compared with a control group consisting of 54 identically defined "high-risk" infants who were on an unrestricted diet. The cumulative prevalence of atopic symptoms was significantly lower at 18 months in the prevention group (32%) than in the control group (74%) ($p < 0.01$) because of a reduced prevalence of recurrent wheezing (13% vs. 37%; $p < 0.01$), atopic dermatitis (14% vs. 31%; $p < 0.01$), vomiting/diarrhea (5% vs. 28%; $p < 0.01$), and infantile colic (9% vs. 24%; p < 0.01). The cumulative prevalence of food allergy was significantly lower in the prevention group (6% vs. 17%; p < 0.05). The authors concluded that feeding high-risk infants with breast milk and/or hypoallergenic formula, combined with the avoidance of solid foods during the first 6 months of life, has a protective effect on the risk of atopic symptoms developing during the first 18 months of life. The American Academy of Pediatrics recommends that mothers of high-risk infants should avoid allergens, such as peanuts and nuts, during lactation [52].

When breastfeeding is not possible, the use of a partially or completely hydrolyzed hypoallergenic formula is desirable for infants at risk [37,52,117,118,121]. A meta-analysis of studies that compared prolonged feeding with hydrolysed formulas versus cow's milk formulas in high risk infants found a significant reduction in infant allergy (seven studies, 2514 infants; typical risk ratio: 0.79; 95% CI: 0.66 to 0.94), but not in the incidence of childhood allergy (two studies, 950 infants; typical risk ratio: 0.85; 95% CI: 0.69 to 1.05 [122].

Partially hydrolysed formulas are often used for the prevention of atopy when breastfeeding is not possible in infants with a strong family history of allergy or elevated cord IgE levels to reduce possible food allergy symptoms [6,14,121]. These formulas have been developed with the aim of minimizing the number of sensitizing epitopes within milk proteins, while at the same time retaining peptides of sufficient size and immunogenicity to stimulate the induction of oral tolerance [37]. Compared with extensively hydrolysed formulas, partially hydrolysed formulas are less expensive and more palatable [46]. Prospective controlled trials examining the use of extensively hydrolysed formulas and partially hydrolysed formulas for allergy prevention among high-risk infants show significant reductions in the cumulative incidence of atopic disease through the first five years of life compared with those fed with cow's milk formulas [121]. In the meta-analysis performed by Osborn et al, infants fed extensively hydrolysed formulas versus partially hydrolysed formulas had a significant reduction in food allergy (two studies, 341 infants; typical risk ratio: 0.43; 95%confidence interval: 0.19 to 0.99), but there was no significant difference in all allergy or any other specific allergy incidence [122].

Soy protein is immunogenic and allergenic, although less than cow's milk [118]. A meta-analysis of two studies (n=283) found no significant differences in childhood allergy cumulative incidence from the use of a soy formula compared to a cow's milk formula (typical risk ratio: 0.67; 95% confidence interval: 0.18 to 2.46) [40]. As there is no evidence of benefit, the use of a soy formula for prevention of food allergy cannot be recommended [40,52].

Early introduction of solid food may increase the risk of food allergy [52,53,102,123,124]. The existing literature suggests that the optimal time for the introduction of selected supplemental food should be six months. For infants at risk, dairy products should not be introduced before 12 months, eggs 24 months, and peanut, tree nuts, fish, and seafood at least 36 months of age [125]. It is recommended that foods should be introduced one at a time and gradually [125]. Cooked, homogenized foods should be preferred to their fresh counterparts when a reduction of allergenicity has been clinically demonstrated for that processed food [125].

It has been suggested that supplementation with long-chain polyunsaturated fatty acids might reduce the incidence of atopic diseases [126-128]. A Cochrane systematic review showed no consistent beneficial effect on marine fatty acids (fish oil) supplementation in asthma prevention [129]. A meta-analysis showed that supplementation with fish oil did not improve the severity of atopic dermatitis [130].

MISCELLANEOUS ASPECTS

Patients with food-dependent exercise-induced anaphylaxis should refrain from exercise within 2 to 3 hours after ingesting the triggering food [66]. Affected patients should exercise with their friends and should stop exercise and seek help immediately if symptoms develop [66].

High-risk environments have to be avoided if the patient is highly allergic and particularly if the patient is allergic to airborne food allergens [5]. High-risk areas include common eating places such as childcare centers, school cafeterias, restaurants, and ice cream shops [5]. School and childcare centers should have policies facilitating food allergen avoidance such as prohibition of sharing of food or utensil and increased staff supervision during meal times [5].

It has been demonstrated that vapors containing proteins emitted from cooking food (e.g. steaming fish) can induce asthmatic attacks [18,67]. Inhalational exposures to foods, particular in the workplace, account for approximately 1% of asthmatic attacks in the adult population [67,131]. Baker's asthma caused by inhalation of flour or mold-derived enzymes used as flour additives is a good example [5]. Affected individuals have asthmatic attacks in association with exposure to aerosolized wheat proteins and have positive skin prick tests or serum specific IgE to wheat proteins [131]. Likewise, peanut dust in airplanes can provoke allergic reactions in susceptible individuals [132]. Allergenic proteins may also reach the respiratory tract via the circulation or they may act via inflammatory mediators released from the skin or gastrointestinal tract [131]. Exposure to known food allergens via inhalation (e.g. flour, peanut dust, steaming fish), skin (e.g. peanut oil in moisturizers), and mucous membrane (e.g. kissing), should be avoided [133].

FUTURE THERAPEUTIC OPTIONS

Currently, there is no effective and safe specific immunotherapy for food allergens. The study of injectable specific immune therapy using peanuts was suspended because of the high rate of adverse reactions [134]. Traditional injection immunotherapy for other food allergies is not recommended either because of the risk of serious systemic reactions associated with such therapy [135]. Sublingual immunotherapy to food allergens is better tolerated and preliminary results are encouraging [48,136,137]. Long-term efficiency of sublingual immunotherapy, however, remains to be determined.

Vaccines for immunotherapy specially for food-induced anaphylaxis that are being developed include humanized anti-IgE monoclonal antibody therapy, sublingual immunotherapy, peptide immunotherapy, mutated protein immunotherapy, plasmid DNA immunotherapy, engineered recombinant protein immunotherapy, immunostimulatory sequence-modulated immunotherapy, cytokine-modulated immunotherapy, bacterial-encapsulated allergen immunotherapy, and homologous protein immunotherapy [2,7,13,137-

140]. Allergen-specific immunotherapy should be considered for patients who have specific IgE antibodies to clinically relevant allergens and whose allergic symptoms are severe enough to warrant the time and risk of allergen immunotherapy [137].

Preliminary studies showed the potential use of humanized monoclonal anti-IgE antibody in food-allergic subjects [141,142]. Humanized monoclonal anti-IgE antibody binds to the third domain of the Fc region of the IgE molecule and prevents its binding to the high affinity receptor on mast cells and basophils [143]. The anti-IgE also downregulates the expression of the high affinity receptor on mast cells and decreases the release of histamine from basophils [144]. In a randomized, double-blind, placebo-controlled trial in 84 patients with a history of peanut allergy, Leung et al showed an increased threshold of tolerance in patients with severe peanut allergy on oral food challenge after being given every 4 weeks subcutaneous injection of TNX-901 for four doses [138]. TNX-901 is a humanized IgG_1 monoclonal antibody against IgE that binds with high affinity to an epitope in the CH_3 domain [138]. The effect was highly significant at the 450 mg dose level. However, even at the highest dose of TNX-901, approximately 25% of patients were not protected. The treatment was well tolerated with no systemic adverse events. Unfortunately, anti-IgE therapy is expensive and such therapy has to be administered on a regular basis so as to maintain its protective effect [35,67]. Currently, another anti-IgE humanized IgG_1 antibody (omalizumab) is being tested in subjects older than 6 years of age with peanut anaphylaxis [19]. Stern et al. treated four adults with eosinophilic esophagitis with a human monoclonal IgG_1 antibody against interleukin-5 (mepolizumab) given by infusion on a monthly basis [145]. After three months of treatment (750 mg monthly), the mean and maximal esophageal eosinophilic count fell from 46 to 6 and from 153 to 28 per high-power field, respectively. The patients also reported improvement of clinical symptoms and quality of life.

Oral immunotherapy seems to represent an interesting and promising approach for the management of food allergy [137,146]. Enrique et al. randomized in a double-blind, placebo-controlled fashion 23 patients with hazelnut allergy to receive either a standardized hazelnut extract or placebo using a sublingual-spit rush protocol over four days [147]. They then received maintenance sublingual immunotherapy for approximately three months. On repeat double-blind, placebo-controlled food challenge, patients in the treatment group had a mean quantity of hazelnut provoking objective symptoms and increased tolerance to hazelnuts from 2.29 gm to 11.59 gm ($p=0.02$) while patients in the placebo group had a non-significant increase from 3.49 gm to 4.14 gm.

Peptide immunotherapy utilizes peptide fragments containing T-cell-reactive epitopes rather complete protein molecules [135]. Apparently, this kind of therapy would induce T-cell unresponsiveness and production of interferon-γ in a concentration-dependent manner [19]. Peptide immunotherapy allows for formulation of vaccines against any target in which major allergenic proteins are known because IgE binding sites for each major allergen do not have to be mapped [19]. It is hoped that such therapy would render T-cells unresponsive to subsequent allergen exposure. Peptide immunotherapy might play a role in the future therapy of food allergy.

Mutated protein immunotherapy is based on the modification of the primary amino acid sequences of IgE-binding allergenic epitopes of the major allergens present in food, with the aim of reducing allergen potential thereby eliminating activation of mast cells and basophils [19,135]. Mutation of the IgE-binding sites leaves the T-cell response unaffected [148]. Such therapy is hampered by the large numbers of allergens present in each food.

Plasmid DNA immunotherapy results in transcription and translation of encoded genes and elicits an antibody response in the host, thereby preferentially induces a Th1 immune response and suppression of IgE [15,149] Plasmid DNA requires immunostimulatory sequences for optimal immunogenicity [149].

Engineered recombinant protein technology makes room for the development of hypoallergenic derivatives of natural allergens, which would minimize the adverse effect of immunotherapy. Such allergens should not be able to activate cells via cross-linking of IgE antibodies, but should preserve T-cell epitopes and activate B-cells to induce blocking IgG antibodies [15].

Immunostimulatory sequence-modulated immunotherapy using CpG has been shown to be effective in reversing IgE-mediated sensitization in patients with ragweed allergy [150,151]. Likewise, immunostimulatory sequence-conjugated Ara h 2 has been shown to be beneficial in the treatment allergy in a murine model [67].

In animal studies, various Chinese herbs have been shown to block anaphylactic reactions resulting from food allergy [152-154]. The therapeutic effect was associated with immunoregulatory effects on Th1-Th-2 responses and reductions in IgE levels [153]. The exact mechanism is not known and further studies are required.

Vaccinations and Food Allergies

Although measles-German measles-mumps (MMR) vaccine is cultured from chick embryos, MMR vaccination is not contraindicated in children allergic to eggs [59,155]. The vaccination should, however, be administered in a supervised setting [59,156]. On the other hand, the vaccine is contraindicated in children with known systemic allergic reaction to neomycin or gelatin [157]. Patients with egg allergy should be tested before getting influenza or yellow fever vaccines, which contain egg protein [158].

Conclusion

Currently, strict avoidance of the allergenic food and ready access to self-injectable epinephrine are the standards of care. Recent advances in our understanding of the immunological processes involved in the pathogenesis of food allergy have opened the door to the investigation and development of novel therapeutic and prophylactic therapies against food allergy. It is hoped that safe and effective vaccines for immunotherapy will be available in the future for the prevention and treatment of food allergy. In particular, anti-IgE therapy may eventually turn out to be a useful adjunct to allergen avoidance.

References

[1] Bangash, SA; Bahna, SL. Pediatric food allergy update. *Curr. Allergy Asthma Rep.* 2005;5:437-444.

[2] Burks, AW; Ballmer-Weber, BK. Food allergies. *Mol. Nutr. Food Res.* 2006;50:595-603.
[3] Chehade, M. IgE and non-IgE-mediated food allergy: treatment in 2007. *Curr. Opin. Allergy Clin. Immunol.* 2007;7:264-268.
[4] Sicherer, SH; Sampson, HA. Food allergy. *J. Allergy Clin. Immunol.* 2006;117:S470-S475.
[5] American College of Allergy, Asthma, & Immunology. Food allergy: a practice parameter. *Ann. Allergy Asthma Immunol.* 2006;96(Suppl 2):S1-S68.
[6] Leung, AK. Food allergy: a clinical approach. *Adv. Pediatr.* 1998;45:145-177.
[7] Scurlock, AM; Lee, LA; Burks, AW. Food allergy in children. *Immunol. Allergy Clin. North Am.* 2005;25:369-388.
[8] Bock, SA. Prospective appraisal of complaints of adverse reactions to foods in children during the first 3 years of life. *Pediatrics.* 1987;79:683-688.
[9] Crespo, JF; Pascual, C; Burks, AW; et al. Frequency of food allergy in a pediatric population from Spain. *Pediatr. Allergy Immunol.* 1995;6:39-43.
[10] Bock, SA; Sampson, HA. Food allergy in infancy. *Pediatr. Clin. North Am.* 1994;41:1047-1067.
[11] Hoffman, KM; Sampson, HA. Evaluation and management of patients with adverse food reactions. In: Bierman, CW; Pearlman, DS; Shapiro, GG; et al. (eds). *Allergy, Asthma, and Immunology from Infancy to Adulthood.* Philadelphia: WB Saunders. 1996; pp.665-686.
[12] Sampson, HA. Food allergies. In: McMillan, JA; Feigin, RD; DeAngelis, C; et al. (eds). *Oski's Pediatrics: Principles & Practice.* Philadelphia: Lippincott Williams & Wilkins. 2006; pp.2417-2423.
[13] Lee, LA; Burks, AW. Food allergies: prevalence, molecular characterization, and treatment/prevention strategies. *Annu. Rev. Nutr.* 2006;26;539-565.
[14] Anderson, JA. Food allergy and intolerance. In: Lieberman, P; Anderson, JA. (ed). *Current Clinical Practice: Allergic Diseases: Diagnosis and Treatment.* 3rd edition. Totowa: Humana Press, 2007, pp. 271-294.
[15] Nieuwenhuizen, N; Lopata, A. Fighting food allergy: current approaches. *Ann. N. Y. Acad. Sci.* 2005;1056:30-45.
[16] Bock, SA; Munoz-Furlong, A; Sampson, HA. Fatalities due to anaphylactic reactions to foods. *J. Allergy Clin. Immunol.* 2001;107:191-193.
[17] Fiocchi, A; Martelli, A. Dietary management of food allergy. *Pediatr. Ann.* 2006;35:755-763.
[18] James, JM; Crespo, JF. Allergic reactions to foods by inhalation. *Curr. Allergy Asthma Rep.* 2007;7:167-174.
[19] Nowak-Wegrzyn, A. Immunotherapy for food allergy. *Inflamm. Allergy Drug Targets.* 2006;5:23-34.
[20] Altman, DR; Chiaramonte, LT. Public perception of food allergy. *J. Allergy Clin. Immunol.* 1996;97:1247-1251.
[21] Opper, PH; Burakoff, R. Food allergy and intolerance. *Gastroenterologist.* 1993;1:211-220.
[22] Sicherer, SH; Leung, DY. Advances in allergic skin disease, anaphylaxis, and hypersensitivity reactions to foods, drugs, and insects. *J. Allergy Clin. Immunol.* 2007;119:1462-1469.

[23] Stern, M. Gastrointestinal allergy. In: Walker, WA; Durie, PR; Hamilton, JR; et al. (eds). *Pediatric Gastrointestinal Disease.* St. Louis: Mosby. 1991; pp.557-574.

[24] Venter, C; Pereira, B; Grundy, J; et al. Incidence of parentally reported and clinically diagnosed food hypersensitivity in the first year of life. *J. Allergy Clin. Immunol.* 2006;117:1118-1124.

[25] Leung AK, Kamat D. Clinical manifestations of food allergy. In: Columbus F, ed. Food Allergies: New Research. New York: Nova Science Publishers, Inc. (in press).

[26] Nash, S; Burks, AW. Oral allergy syndrome. *Curr. Allergy Asthma Rep.* 2007;7:1-2.

[27] Grimshaw, KE. Dietary management of food allergy in children. *Proc. Nutr. Soc.* 2006;65:412-417.

[28] Joint Task Force on Practice Parameters; American Academy of Allergy, Asthma and Immunology; American College of Allergy, Asthma and Immunology; Joint Council of Allergy, Asthma and Immunology. The diagnosis and management of anaphylaxis: an updated practice parameter. *J. Allergy Clin. Immunol.* 2005;115:S483-S523.

[29] Muñoz-Furlong, A. Daily coping strategies for patients and their families. *Pediatrics.* 2003;111:1654-1661.

[30] Joshi, P; Mofidi, S; Sicherer, SH. Interpretation of commercial food ingredient labels by parents of food-allergic children. *J. Allergy Clin. Immunol.* 2002;109:1019-1021.

[31] Taylor, SL; Hefle, SL. Food allergen labeling in the USA and Europe. *Curr. Opin. Allergy Clin. Immunol.* 2006;6:186-190.

[32] European Commission. Directive 2003/89/EC of the European Parliament and of the Council of 10 November 2003 amending Directive 2000/13/EC as regards indication of the ingredients present in foodstuffs. *Official J. Euro. Comminities.* 2003;L308:15-18.

[33] Lee, BW. Food and allergy. *Ann. Acad. Med. Singapore.* 1995;24:238-241.

[34] Steinman, HA. "Hidden" allergens in foods. *J. Allergy Clin. Immunol.* 1996;98:241-250.

[35] Keet, CA; Wood RA. Food allergy and anaphylaxis. *Immunol. Allergy Clin. North Am.* 2007;27:193-212.

[36] Leung, AK; Sauve, RS. Whole cow's milk in infancy. *Paediatr. Child. Health.* 2003;8:419-421.

[37] Crittenden, RG; Bennett, LE. Cow's milk allergy: a complex disorder. *J. Am. Coll. Nutr.* 2005;24:S582-S591.

[38] Garcia-Careaga, M; Kerner, JA, Jr. Gastrointestinal manifestations of food allergies in pediatric patients. *Nutr. Clin. Pract.* 2005;20:526-535.

[39] Duggan, C; Walker, WA. Protein intolerance. In: Oski, FA; DeAngelis, CD; Feigin, RD. (eds). *Principles and Practice of Pediatrics.* Philadelphia: JB Lippincott. 1994; pp.1887-1890.

[40] Osborn, DA; Sinn, J. Soy formula for prevention of allergy and food intolerance in infants (review). *Cochrane Database Syst. Rev.* 2006 Oct 18;(4):CD003741.

[41] Bellioni-Businco, B; Paganelli, R; Lucenti, P; et al. Allergenicity of goat's milk in children with cow's milk allergy. *J. Allergy Clin. Immunol.* 1999;103:1191-1194.

[42] Beyer, K. Hypoallergenicity: a principle for the treatment of food allergy. *Nestlé Nutr. Workshop Ser. Pediatr. Program.* 2007;59:37-47.

[43] Committee on Nutrition, American Academy of Pediatrics. Food hypersensitivity. In: Barness, LA .(ed). *Pediatric Nutrition Handbook.* Elk Grove Village, Ill: American Academy of Pediatrics. 1993; pp.274-285.

[44] Sampson, HA; Bernhisel-Broadbent, J; Yang, E; et al. Safety of casein hydrolysate formula in children with cow milk allergy. *J. Pediatr.* 1991;118:520-525.

[45] Sampson, HA; James, JM; Bernhisel-Broadbent, J. Safety of an amino acid-derived infant formula in children allergic to cow's milk. *Pediatrics.* 1992;90:463-465.

[46] Leung, AK; Bowen, TJ. Seasonal allergic rhinitis and food allergy. In: Bergman, AB. (ed). *Twenty Common Problems in Pediatrics.* New York: McGraw-Hill. 2001, pp. 219-233.

[47] Stern, M. Allergic enteropathy. In: Walker, WA; Durie, PR; Hamilton, JR; et al. (eds). *Pediatric Gastrointestinal Disease.* St. Louis: Mosby. 1996; pp.677-692.

[48] Chehade, M; Magid, MS; Mofidi, S; et al. Allergic eosinophilic gastroenteritis with protein-losing enteropathy: intestinal pathology, clinical course, and long-term follow-up. *J. Pediatr. Gastroenterol. Nutr.* 2006;42:516-521.

[49] Kagalwalla, AF; Sentongo, TA; Ritz, S; et al. Effect of six-food elimination diet on clinical and histologic outcomes in eosinophilic esophagitis. *Clin. Gastroenterol. Hepatol.* 2006;4:1097-1102.

[50] Liacouras, CA. Eosinophilic esophagitis: treatment in 2005. *Curr. Opin. Gastroenterol.* 2006;22:147-142.

[51] Miller TP, Leung AK. Eosinophilic gastrointestinal disorders. In: Columbus F, ed. Food Allergies: New Research. New York: Nova Science Publishers, Inc. (in press).

[52] American Academy of Pediatrics, Committee on Nutrition. Hypoallergenic infant formulas. *Pediatrics.* 2000;106:346-349.

[53] Hyams, JS. Food allergy (food hypersensitivity). In: Kliegman, RM; Behrman, RE; Jenson, HB; et al. *Nelson Textbook of Pediatrics.* Philadelphia: Saunders Elsevier. 2007, pp.1585-1587.

[54] Burks, AW; Sampson, H. Food allergies in children. *Curr. Probl. Pediatr.* 1993;23:230-252.

[55] Dannaeus, A. Food allergy in infancy and children. *Ann. Allergy.* 1987;59:124-126.

[56] Tariq, SM; Stevens, M; Matthew, S; et al. Cohort study of peanut and tree nut sensitisation by age of 4 years. *BMJ.* 1996;313:514-517.

[57] Golbert, TM. Food allergy and immunological diseases of the gastrointestinal tract. In: Patterson, R; Zeiss, CR; Grammer, LC. (eds). *Allergic Diseases - Diagnosis and Management.* Philadelphia: JB Lippincott. 1993; pp.353-394.

[58] Metcalfe, DD. Allergic reactions to foods. In: Frank, MM; Austen, KF; Claman, HN; et al.(eds). *Samter's Immunological Diseases.* Boston: Little, Brown. 1995; pp.1357-1366.

[59] Baral, VR; Hourihane, JO. Food allergy in children. *Postgrad. Med. J.* 2005;81:693-701.

[60] Muraro, A; Roberts, G; Clark, A; et al. The management of anaphylaxis in childhood: position paper of the European Academy of Allergology and Clinical Immunology. *Allergy.* 2007;62:857-871.

[61] Wang, J; Sampson, HA. Food anaphylaxis. *Clin. Exp. Allergy.* 2007;37:651-660.

[62] Simons, FE; Roberts, JR; Gu, X; et al. Epinephrine absorption in children with a history of anaphylaxis. *J. Allergy Clin. Immunol.* 1998;101:33-37.

[63] Brown, SG. Cardiovascular aspects of anaphylaxis: implications for treatment and diagnosis. *Curr. Opin. Allergy Clin. Immunol.* 2005;5:359-364.

[64] Oren, E; Banerji, A; Clark, S; et al. Food-induced anaphylaxis and repeated epinephrine treatments. *Ann. Allergy Asthma Immunol.* 2007;99:429-432.
[65] Simons, FE. First-aid treatment of anaphylaxis to food: focus on epinephrine. *J. Allergy Clin. Immunol.* 2004;113:837-844.
[66] Sampson, HA; Leung, DY. Anaphylaxis. In: Kliegman, RM; Behrman, RE; Jenson, HB; et al. (eds). *Nelson Textbook of Pediatrics*. Philadelphia: Saunders Elsevier. 2007, pp.983-985.
[67] Sampson, HA. Update on food allergy. *J. Allergy Clin. Immunol.* 2004;113:805-819.
[68] Kelso, JM. A second dose of epinephrine for anaphylaxis: how often needed and how to carry. *J. Allergy Clin. Immunol.* 2006;117:464-465.
[69] Clark, S; Camargo, CA, Jr. Emergency management of food allergy: systems perspective. *Curr. Opin. Allergy Clin. Immunol.* 2005;5:293-298.
[70] Du Toit, G; Fox, A; Morris, A. Managing food allergy in children. *Practitioner.* 2006;250:45-46, 49-52.
[71] Metcalfe, DD. Food hypersensitivity. *J. Allergy Clin. Immunol.* 1984;73:749-762.
[72] Leung, AK; Hon, KL; Robson, WL. Atopic dermatitis. *Adv. Pediatr.* 2007;54:241-273.
[73] Bailey, M; Haverson, K; Inman, C; et al. The development of the mucosal immune system pre- and post-weaning: balancing regulatory and effector function. *Proc. Nutr. Soc.* 2005;64:451-457.
[74] Kalliomäki, M; Kirjavainen, P; Eerola, E; et al. Distinct patterns of neonatal gut microflora in infants in whom atopy was and was not developing. *J. Allergy Clin. Immunol.* 2001;107:129-134.
[75] Watanabe, S; Narisawa, Y; Arase, S; et al. Differences in fecal microflora between patients with atopic dermatitis and healthy control subjects. *J. Allergy Clin. Immunol.* 2003;111:587-591.
[76] Björkstén, B; Sepp, E; Julge,K; et al. Allergy development and the intestinal microflora during the first year of life. *J. Allergy Clin. Immunol.* 2001;108:516-520.
[77] Laitenen, K; Isolauri, E. Management of food allergy: vitamins, fatty acids or probiotics? *Eur. J. Gastroenterol. Hepatol.* 2005;17:1305-1311.
[78] Boyle, RJ; Tang, ML. The role of probiotics in the management of allergic disease. *Clin. Exp. Allergy.* 2006;36:568-576.
[79] He, F; Morita, H; Hashimoto, H; et al. Intestinal *Bifidobacterium* species induce varying cytokine production. *J. Allergy Clin. Immunol.* 2002;109:1035-1036.
[80] Kirjavainen, PV; Apostolou, E; Salminen, SJ, et al. New aspects of probiotics – a novel approach in the management of food allergy. *Allergy.* 1999;54:909-915.
[81] O'Sullivan, GC; Kelly, P; O'Halloran, S; et al. Probiotics: an emerging therapy. *Curr. Pharm. Des.* 2005;11:3-10.
[82] Rosenfeld, V; Benfeldt, E; Valerius, NH; et al. Effect of probiotics on gastrointestinal symptoms and small intestinal permeability in children with atopic dermatitis. *J. Pediatr.* 2004;145:612-616.
[83] Williams, HC. Two "positive" studies of probiotics for atopic dermatitis – or are they? *Arch. Dermatol.* 2006;142:1201-1203.
[84] Kalliomäki, M; Salminen, S; Arvilommi, H; et al. Probiotics in primary prevention of atopic disease: a randomized placebo-controlled trial. *Lancet.* 2001b;357:1076-1079.

[85] Kalliomäki, M; Salminen, S; Poussa, T; et al. Probiotics and prevention of atopic disease: 4-year follow-up of a randomized placebo-controlled trial. *Lancet.* 2003;361:1869-1871.

[86] Lodinová-Zádníková, R; Cukrowska, B; Tlaskalova-Hogenova, H. Oral administration of probiotic *Escherichia coli* after birth reduces frequency of allergies and repeated infections later in life (after 10 and 20 years). *Int. Arch. Allergy Immunol.* 2003;131:209-211.

[87] Rosenfeld, V; Benfeldt, E; Nielsen, SD; et al. Effect of probiotic *lactobacilli* strains in children with atopic dermatitis. *J. Allergy Clin. Immunol.* 2003;111:389-395.

[88] Viljanen, M; Savilahti, E; Haahtela, T; et al. Probiotics in the treatment of atopic eczema/dermatitis syndrome in infants: a double-blind placebo-controlled trial. *Allergy.* 2005;60:494-500.

[89] Tamura, M; Shikina, T; Morihana, T; et al. Effects of probiotics on allergic rhinitis induced by Japanese cedar pollen: randomized double-blind, placebo-controlled clinical trial. *Int. Arch. Allergy Immunol.* 2007;143:75-82.

[90] Taylor, AL; Dunstan, JA; Prescott, SL, et al. Probiotic supplementation for the first 6 months of life fails to reduce the risk of atopic dermatitis and increases the risk of allergen sensitization in high-risk children: a randomized controlled trial. *J. Allergy Clin. Immunol.* 2007;119:184-191.

[91] Kukkonen, K; Savilahti, E; Haahtela, T; et al. Probiotic and prebiotic galacto-oligosaccharides in the prevention of allergic diseases: a randomized, double-blind, placebo-controlled trial. *J. Allergy Clin. Immunol.* 2007;119:192-198.

[92] Weston, S; Halbert, A; Richmond, P; et al. Effects of probiotics on atopic dermatitis: a randomized controlled trial. *Arch. Dis. Child.* 2005;90:892-897.

[93] Passeron, T; Lacour, JP; Fontas, E; et al. Prebiotics and synbiotics: two promising approaches for the treatment of atopic dermatitis in children above 2 years. *Allergy.* 2006;61:431-437.

[94] Chulada, PC; Arbes, SJ Jr; Dunson, D; et al. Breast-feeding and the prevalence of asthma and wheeze in children: analyses from the Third National Health and Nutrition Examination Survey, 1988-1994. *J. Allergy Clin. Immunol.* 2003;111:328-336.

[95] Gdalevich, M; Mimouni, D; Mimouni, M. Breast-feeding and the risk of bronchial asthma in childhood: a systematic review with meta-analysis of prospective studies. *J. Pediatr.* 2001;139:261-266.

[96] Leung, AK; Barber, KA. Managing childhood atopic dermatitis. *Adv. Ther.* 2003;20:120-137.

[97] Leung, AK; Sauve, RS. Breast is best for babies. *J. Natl. Med. Assoc.* 2005;97:1010-1019.

[98] Hanifin, JM; Cooper, KD; Ho, VC; et al. Guidelines for care of atopic dermatitis. *J. Am. Acad. Dermatol.* 2004;50:391-404.

[99] Muraro, A; Dreborg, S; Halken, S; et al. Dietary prevention of allergic diseases in infants and children. Part III: critical review of published peer-reviewed observational and interventional studies and final recommendations. *Pediatr. Allergy Immunol.* 2004;15:291-307.

[100] Kull, I; Böhme, M; Wahlgren, CF; et al. Breast-feeding reduces the risk for childhood eczema. *J. Allergy Clin. Immunol.* 2006;116:657-661.

[101] Mimouni Bloch, AM; Mimouni, D; Mimouni, M; et al. Does breastfeeding protect

against allergic rhinitis during childhood? A meta-analysis of prospective studies. *Acta Paediatr.* 2002;91:275-279.

[102] Siltanen, M; Kajosaari, M; Poussa, T; et al. A dual long-term effect of breastfeeding on atopy in relation to heredity in children at 4 years of age. *Allergy.* 2003;58:524-530.

[103] Host, A; Halken, S. Primary prevention of food allergy in infants who are at risk. *Curr. Opin. Allergy Clin. Immunol.* 2005;5:255-259.

[104] Schmitz, J. Prevention of food allergy during late infancy and early childhood. *Nestle Nutr. Workshop Ser. Pediatr. Program.* 2005;56:15-25.

[105] Sampson, HA; Fomon, SJ. Antigen-antibody interactions and adverse reactions to food. In: Fomon, SJ. (ed). *Nutrition of Normal Infants.* St. Louis: Mosby. 1993; pp.395-408.

[106] Zeiger, RS; Heller, S; Mellon, MH; et al. Effect of combined maternal and infant food-allergen avoidance on development of atopy in early infancy: a randomized study. *J. Allergy Clin. Immunol.* 1989;84:72-89.

[107] Brostoff, J; Hawk, LJ. Food allergy in children. *Eur. J. Clin. Nutr.* 1991;45:S11-S15.

[108] Leung, AK; Robson, WL. Breastfeeding. In: Carter, LV. (ed). *Child Nutrition Research Advances.* New York: Nova Science Publishers, Inc. (in press).

[109] Björkstén, B; Kjellman, NI. Does breast-feeding prevent food allergy? Allergy Proc. 1991;12:233-237.

[110] Bahna, SL. Management of food allergies. *Ann. Allergy.* 1984;53:678-682.

[111] Kramer, MS; Kakuma, R. Maternal dietary antigen avoidance during pregnancy and/or lactation for preventing or treating atopic disease in the child. *Cochrane Database Syst. Rev.* 2003(4):CD000133.

[112] Chandra, RK; Puri, S; Hamed, A. Influence of maternal diet during lactation and use of formula feeds on development of atopic eczema in high risk infants. *BMJ.* 1989;299:228-230.

[113] Hattevig, G; Sigurs, N; Kjellman, B. Maternal food antigen avoidance during lactation and allergy during the first 10 years of age. *J. Allergy Clin. Immunol.* 1996;97:241.

[114] Chandra, RK; Puri, S; Suraiya, C; et al. Influence of maternal food antigen avoidance during pregnancy and lactation on incidence of atopic eczema in infants. *Clin. Allergy.* 1986;16:563-571.

[115] Hattevig, G; Sigurs, N; Kjellman, B. Effects of maternal dietary avoidance during lactation on allergy in children at 10 years of age. *Acta Paediatr.* 1999;88:7-12.

[116] Falth-Magnusson, K; Kjellman, NIM. Development of atopic disease in babies whose mothers were receiving exclusion diet during pregnancy: a randomized study. *J. Allergy Clin. Immunol.* 1987;80:868-875.

[117] von Berg, A. The concept of hypoallergenicity for atopy prevention. *Nestlé Nutr. Workshop Ser. Pediatr. Program.* 2007;59:49-62.

[118] Zeiger, RS. Food allergen avoidance in the prevention of food allergy in infancy and children. *Pediatrics.* 2003;111:1662-1671.

[119] Marini, A; Agosti, M; Motta, G; et al. Effects of a dietary and environmental prevention programme on the incidence of allergic symptoms in high atopic risk infants: three years follow-up. *Acta Paediatr.* 1996;414:S1-S22.

[120] Halken, S; Host, A; Hansen, LG; et al. Effect of an allergy prevention programme on incidence of atopic symptoms in infancy: a prospective study of 159 "high-risk" infants. *Allergy.* 1992;47:545-553.

[121] Hays, T; Wood, RA. A systematic review of the role of hydrolysed infant formulas in allergy prevention. *Arch. Pediatr. Adolesc. Med.* 2005;159:810-816.
[122] Osborn, DA; Sinn, J. Formulas containing hydrolysed protein for prevention of allergy and food intolerance in infants (review). *Cochrane Database Syst. Rev.* 2006 Oct 18;(4):CD003664.
[123] Kajosaari, M; Saarinen, UM. Prophylaxis of atopic disease by six months' total solid foods elimination. *Acta Paediatr. Scand.* 1983;72:411-421.
[124] Morgan, J; Williams, P; Norris, F; et al. Eczema and early solid feeding in preterm infants. *Arch. Dis. Child.* 2004;89:309-314.
[125] Fiocchi, A; Assa'ad, A; Bahna, S; et al. Food allergy and the introduction of solid foods to infants: a consensus document. *Ann. Allergy Asthma Immunol.* 2006;97:10-21.
[126] [126] Hodge, L; Salome, CM; Peat, JK; et al. Consumption of oily fish and childhood asthma risk. *Med. J. Aust.* 1998;164:137-140.
[127] Mayser, P; Mayer, K; Mahloudjian, M; et al. A double-blind, randomized, placebo-controlled trial of n-3 versus n-6 fatty acid-based lipid infusion in atopic dermatitis. *J. Parent. Enter. Nutr.* 2002;26:151-158.
[128] Nafstad, P; Nystad, W; Magnus, P; et al. Asthma and allergic rhinitis at 4 years of age in relation to fish consumption in infancy. *J. Asthma*. 2003;40:343-348.
[129] Thien, FCK; Woods, R; de Lucas S; et al. Dietary marine fatty acids (fish oil) for asthma in adults and children. *Cochrane Database Sys. Rev.* 2002;3:CD001283.
[130] Gool, CJ; Zeegers, MP; Thijs, C. Oral essential fatty acid supplementation in atopic dermatitis – a meta-analysis of placebo-controlled trials. *Br. J. Dermatol.* 2004;150:728-740.
[131] Roberts, G; Lack, G. Relevance of inhalational exposure to food allergens. *Curr. Opin. Allergy Clin. Immunol.* 2003;3:211-215.
[132] Sicherer, SH; Furlong, TJ; DeSimone, J; et al. Self-reported allergic reactions to peanuts on commercial airlines. *J. Allergy Clin. Immunol.* 1999;104:186-189.
[133] Nowak-Wegrzyn, A. Food allergy to proteins. *Nestlé Nutr. Workshop Ser. Pediatr. Program*. 2007;59:17-35.
[134] Nelson, HS; Lahr, J; Rule, R, et al. Treatment of anaphylactic sensitivity to peanuts by immunotherapy with injections of aqueous peanuts extract. *J. Allergy Clin. Immunol.* 1997;99:744-751.
[135] Pons, L; Burks, W. Novel treatments for food allergy. *Expert Opin. Investig. Drugs*. 2005;14:829-834.
[136] Allen, KJ; Hill, DJ, Heine, RG. 4. Food allergy in childhood. *MJA*. 2006;185:394-400.
[137] Enrique, E; Cisteró-Bahíma, A. Specific immunotherapy for food allergy: basic principles and clinical aspects. *Curr. Opin. Allergy Clin. Immunol.* 2006;6:466-469.
[138] Leung, DY; Sampson, HA; Yunginger, JW; et al. Effect of anti-IgE therapy in patients with peanut allergy. *N. Engl. J. Med.* 2003;348:986-993.
[139] Li, XM; Zhang, TF; Huang, CK; et al. Food Allergy Herbal Formula-1 (FAHF-1) blocks peanut-induced anaphylaxis in a murine model. *J. Allergy Clin. Immunol.* 2001;108:639-646.
[140] Li, XM; Srivastava, K; Grishin, A; et al. Persistent protective effect of heat-killed *Escherichia coli* producing "engineered," recombinant peanut proteins in a murine

model of peanut allergy. *J. Allergy Clin. Immunol.* 2003;112:159-167.

[141] Mankad, VS; Burks, AW. Omalizumab: other indications and unanswered questions. *Clin. Rev. Allergy Immunol.* 2005;29:17-30.

[142] Okubo, K; Ogino, S; Nagakura, T; et al. Omalizumab is effective and safe in the treatment of Japanese cedar pollen-induced seasonal allergic rhinitis. *Allergol. Int.* 2006;55:379-386.

[143] Crespo, JF; James, JM; Rodriguez J. Diagnosis and therapy of food allergy. *Mol. Nutr. Food Res.* 2004;48:347-355.

[144] MacGlashan, DW; Bochner, BS; Adelman, DC, et al. Down-regulation of Fc (epsilon) RI expression on human basophils during vivo treatment of atopic patients with anti-IgE antibody. *J. Immunol.* 1997;158:1438-1445.

[145] Stein, ML; Collins, MH; Villanueva, JM; et al. Anti-IL-5 (mepolizumab) therapy for eosinophilic esophagitis. *J. Allergy Clin. Immunol.* 2006;118:1312-1319.

[146] Bieber, T. Allergen-specific sublingual immunotherapy: less mystic, more scientific. *Allergy.* 2006;61:149-150.

[147] Enrique, E; Pineda, F; Malek, T; et al. Sublingual immunotherapy for hazelnut food allergy: a randomized double-blind placebo-controlled study with a standardized hazelnut extract. *J. Allergy Clin. Immunol.* 2005;116:1073-1079.

[148] King, N; Helm, R; Stanley, JS; et al. Allergenic characteristics of a modified peanut allergen. *Mol. Nutr. Food Res.* 2005;49:963-971.

[149] Peng, HJ; Su, SN; Chang, ZN; et al. Induction of specific Th1 responses and suppression of IgE antibody formation by vaccination with plasmid DNA encoding Der f 11. *Vaccine.* 2002;20:1761-1768.

[150] Horn, AA; Raz, E. Immunostimulatory sequence oligodeoxynucleotide-based vaccination and immunomodulation: two unique but complementary strategies for the treatment of allergic diseases. *J. Allergy Clin. Immunol.* 2002;110:706-712.

[151] Marshall, JD; Abtahi, S; Eiden, JJ; et al. Immunostimulatory sequence DNA linked to the Amb a 1 allergen promotes T(H)1 cytokine expression while downregulating T(H)2 cytokine expression in PBMCs from human patients with ragweed allergy. *J. Allergy Clin. Immunol.* 2001;108:191-197.

[152] Hsieh, KY; Hus, CI; Lim JY; et al. Oral administration of an edible-mushroom-derived protein inhibits the development of food-allergic reactions in mice. *Clin. Exp. Allergy.* 2003;33:1595-1602.

[153] Li, XM; Srivastava, K; Huleatt, JW; et al. Engineered recombinant peanut proteins and heat-killed *Listeria monocytogenes* coadministration protects against peanut-induced anaphylaxis in a murine model. *J. Immunol.* 2003;170:3289-3295.

[154] Srivastava, KD; Kattan, JD; Zou, ZM; et al. The Chinese herbal medicine formula FAHF-2 completely blocks anaphylactic reactions in a murine model of peanut allergy. *J. Allergy Clin. Immunol.* 2005;115:171-178.

[155] James, JM; Burks, AW; Robertson, PK; et al. Safe administration of the measles vaccine to children allergic to eggs. *N. Engl. J. Med.* 1995;332:1262-1266.

[156] Khakoo, GA; Lack, G. Recommendations for using MMR vaccine in children allergic to eggs. *BMJ.* 2000;320:929-932.

[157] Lacksman, R; Finn, A. MMR vaccine and allergy. *Arch. Dis. Child.* 2000;82:93-95.

[158] Sampson, HA; Leung, DY. Adverse reactions to food. In: Kliegman, RM; Behrman, RE; Jenson, HB; et al. (eds). *Nelson Textbook of Pediatrics.* Philadelphia: Saunders Elsevier. 2007, pp.986-990.

In: Food Allergies: New Research
Editor: Carrie M. Chesterton, pp. 157-172
ISBN 978-1-60456-978-0

Chapter 5

INFANTILE COLIC: AN UPDATE*

***Alexander K.C. Leung*+**
University of Calgary, Alberta Children's Hospital,
Calgary, Alberta, T2M 0H5, Canada

ABSTRACT

Infantile colic is characterized by paroxysms of uncontrollable crying or fussing in an otherwise healthy and well-fed infant less than 3 months of age. The duration of crying is more than 3 hours per day and more than 3 days per week for at least 3 weeks.

The condition can be very stressful to the family. The etiology is multifactorial. There is increasing evidence that cow's milk proteins may play an important role in the pathogenesis of infantile colic in a significant number of cases. Also, maternal ingestion of eggs, chocolate, citrus fruits, nuts, as well as certain seafood whilst breastfeeding may result in infantile colic. Intestinal permeability to macromolecules, a mechanism of acquired food allergy appears to be increased in some infants with colic. Supportive counseling, reassurance, and dietary modifications (if indicated) are the core measures used for the management of this condition. Use of hypoallergenic diets by breasting mothers should be considered at least for those infants with severe colic or with atopic features such as atopic dermatitis, asthma, and allergic rhinitis. For formula-fed infants with mild to moderate colic, the present consensus is that changing to another formula is usually not necessary. Formula-fed infants with severe colic, especially those with atopic features or a strong family history of atopy may have a beneficial effect from hypoallergenic formulas such as whey hydrolysates or casein hydrolysates. Periodic challenges at monthly intervals are necessary to ensure that the improvement is related to dietary modification and not a result of natural resolution. In most infants, infantile colic resolves by 3 to 4 months of age.

Keywords: infantile colic, cow's milk protein, hypoallergenic diet or formula, reassurance.

---

* Portions of this article appeared previously in *Paediatrics & Child Health* [1] and the *Journal of the Royal Society of Health* [2], and have been used with permission from the Canadian Paediatric Society and The Royal Society for the Promotion of Health.

+ Correspondence to: Dr. Alexander K.C. Leung, #200, 233 - 16th Avenue NW, Calgary, Alberta T2M 0H5; Telefax: (403) 230-3322, E-mail: aleung@ucalgary.ca

Introduction

Infantile colic is generally defined as paroxysms of uncontrollable crying or fussing in an otherwise healthy and well-fed infant less than 3 months of age; the duration of crying is more than 3 hours per day and more than 3 days per week for at least 3 weeks [1-3]. The condition can be profoundly disturbing both to the infant and family [4]. It is one of the most common problems encountered by primary care physicians and child care providers. There is increasing evidence that dietary modification may have a role to play in the management of infantile colic [5,6]. An updated review of infantile colic is therefore in order and is the purpose of the present chapter.

Epidemiology

Prevalence rates in prospective studies varied from 3 to 28% and in retrospective studies from 8 to 40% [6-10]. Prospective studies are more reliable as retrospective studies are prone to recall bias. The two best prospective studies yielded prevalence rates of 5% and 19%, respectively, in the Caucasian population [7,10]. The prevalence rate in the Asian population is considerably lower. In a prospective study of 160 Korean infants, no case of infantile colic was found [11]. The wide variation in the prevalence rate may be explained by the differences in the study design, method of data collection, site of recruitment, and definition of infantile colic [12,13].

Colic occurs equally in both male and female infants [4,14]. Infants with colic tend to have siblings who also have this condition [15]. While some investigators have reported an increased incidence in first-born infants [9,16], others have not found such an association [17,18]. Some investigators have found that the condition is more common in low birth weight infants [19]; others disagree [9]. Seasonal variations have not been implicated in the pervasiveness of infantile colic [20].

Etiology and Pathogenesis

The exact etiology and pathogenesis of infantile colic are not completely understood. The condition is likely multifactorial [15].

Food Allergy

There is increasing evidence that cow's milk proteins may play an important role in the pathogenesis of infantile colic [6,21-25]. Approximately 25% of infants with moderate or severe colic have allergy to cow's milk protein [26,27]. Lothe and Lindberg showed that colic disappeared in 24 of 27 infants when they were given a cow's milk-free diet [24]. These infants were entered into a double-blind placebo-controlled crossover trial of whey protein.

Eighteen infants receiving the whey protein capsules reacted with colic, two infants received placebo reacted with colic, and four infants did not react at all.

Iacono et al. put 70 cow's milk formula-fed infants with severe colic on soy-milk formula. In 50 infants, there was a remission of symptoms when cow's milk protein was eliminated from their diet [23]. Two successive challenges caused the return of symptoms in all these 50 infants. Follow-ups, after an average period of 18 months, showed that in 22 of 50 (44%) of the infants who had cow's milk protein-related colic and 1 to 20 (5%) of those with non-cow's milk protein-related colic developed an overt form of alimentary intolerance. Lucassen et al. randomly selected 43 healthy infants with colic to receive whey hydrolysate formula or standard formula [6]. They found a decrease in the duration of crying in those infants fed with whey hydrolysate formula.

Jakobsson et al. studied the effectiveness of 2 formulae with extensively hydrolysed casein in 22 infants with severe colic [5]. One infant was considered as treatment failure and six infants as protocol failures. The remaining 15 infants showed a significant decrease in the lengths of time they cried as well as a decrease in the intensity of their crying on both formulae. When the infants were challenged in a double-blind design, 11 infants reacted with an increase in crying time to cow's milk protein or bovine whey protein.

Hill et al. studied the effect of diet change in 38 bottle-fed and 77 breast-fed colicky infants in a double-blind, randomized, placebo-controlled trial [22]. Bottle-fed infants were assigned to either casein hydrolysate or cow's milk formula. All mothers of breast-fed infants were started on an artificial color-free, preservative-free, additive-free diet and were randomized to receive either an active low allergen (milk free) diet or a control diet. Hill et al showed that infants on the active diet had their distress reduced by 39% compared with 16% for those on the control diet [22].

Jakobsson and Lindberg put 66 mothers of 66 breast-fed infants with infantile colic on a cow's milk-free diet [28]. The colic disappeared in 35 infants; it reappeared after reintroduction of cow's milk into the mother's diet in 23 of the 35 infants. A double-blind crossover trial with cow's milk whey protein was performed in 16 of these 23 mothers and infants. Six infants had to be taken out of the study for various reasons. Of the remaining 10 infants, nine displayed signs of colic after their mothers had taken the whey-filled capsules.

Maternal ingestion of eggs, chocolate, citrus fruits, nuts, as well as certain seafood whilst breastfeeding may result in infantile colic [29,30]. Hill et al. randomized mothers of 107 term breastfed infants younger than 6 weeks of age with colic to follow a low-allergen diet with elimination of dairy products, soy, wheat, eggs, peanuts, tree nuts, and fish (n=53) and a control group (n=540) whose diet contained the known allergen [31]. Forty seven mothers in the treatment group and 43 mothers in the control group completed the study. Infants were identified as responders if there was at least 25% reduction in duration of crying/fussing on days 8 and 9. The authors showed that 74% of infants in the treatment group versus 37% of infants in the control group were responders (p=<0.001).

Lack of Breastfeeding

Breastfed infants are less likely to suffer from infantile colic [2,32]. For those breastfed infants who have colic, there is a positive correlation with maternal consumption of cow's milk and allergenic products [1,22].

Flatulence

Some physicians believe that excessive amount of intestinal gas causes abdominal distention and intestinal spasm, often leading to infantile colic [18]. Indeed, the word "colic" is derived from the Greek work *kolikos*, the adjective of *kolon* (the large intestine). Flatulence results mainly from excessive air intake or inadequate burping after feeding [15]. Several researchers have reported that colicky infants produce more breath hydrogen in the fasting state and in response to feedings containing lactose than non-colicky infants [33,34]. The increased breath hydrogen excretion may represent an increased lactose malabsorption, differences in colonic bacterial fermentation conditions, or differences in handling of colonic gas [34]. Some authors suggest that lactose intolerance due to a relative deficiency of lactase in the first few months of life is a causative factor in infantile colic [35,36]. Fermentation of lactose by the intestinal flora leads to the production of hydrogen and lactic acid. Lactic acid and lactose draw water into the colon through the process of osmosis. The influx of water into the colon together with hydrogen production results in colonic distension. Hyams et al., however, found no difference in breath hydrogen levels between colicky and non-colicky infants [37]. Studies examining the effect of removing lactose from an infant's diet or adding lactase to break down lactose yielded inconsistent results [35,38,39]. Kanabar et al. found a significant difference in crying time and breath hydrogen in those infants who used the lactase-treated feed [35]. Other authors found that removing lactose from or adding lactase to the feed did not result in the elimination of symptoms of colic [38,39].

Intestinal Hormone Abnormalities

Intestinal permeability to macromolecules, a mechanism of acquired food allergy appears to be increased in some infants with colic [40]. It is possible that the increased permeability to macromolecules reflects an immature function of the gastrointestinal tract.

Intestinal spasm is another suggested cause of colic. Motilin stimulates gastric and intestinal motility and high motility levels may lead to intestinal spasms. In a prospective study, Lothe et al. found increased serum levels of motilin in cord blood and in blood from neonates who later developed colic [41]. This finding might indicate that the intestinal tract is affected before any symptoms of colic appear. Also, exposure to cigarette smoke is associated with increased plasma and intestinal motilin level which may result in infantile colic [42].

Using ultrasonography, Lehtonen et al have shown decreased contractility of the gallbladder in colicky infants [43]. The hypocontractility of the gallbladder may be due to a disturbance in cholecystokinin secretion.

Kurtoglu et al. measured urinary levels of 5-hydroxy-3-indole acetic acid, a metabolite of serotonin, in 16 infants with infantile colic and 10 control infants [44]. These authors found high urinary 5-hydroxy-3-indole acetic acid levels in colicky infants and suggested that high serotonin levels might be responsible for the colic. Although some studies suggest that serum hormones may be markers for the variability in physiological function of the intestinal tract [41,44], their roles as determinants of infantile colic remain elusive.

Recently, it has been shown that colicky infants have higher levels of ghrelin compared to non-colicky infants [45]. Ghrelin has been implicated in the causation of hyperperistalsis and infantile colic [45].

Intestinal Microflora Imbalance

It has been shown that colicky infants have lower counts of intestinal lactobacilli compared to non-colicky infants [46,47]. An imbalance of intestinal microflora might lead to aberrant antigen transfer across the gut barrier and increased vulnerability to the breakdown of oral tolerance [25,37]. Further, *Lactobacillus brevis* and *L. lactis lactis* have been implicated in the pathogenesis of infantile colic by increasing gas production and abdominal distension [25].

Dysregulation of the Nervous System

Transient developmental dysregulation of the nervous system resulting in intestinal hypermotility has been suggested as a cause of infantile colic [48]. Infants with a low threshold for over-stimulation may respond to environmental stimulation with excessive irritability and crying. As the maturing central nervous system becomes less sensitive to external stimulation, the colicky symptoms are also alleviated [15].

Kirjavainen et al. performed heart rate variability analysis on 12 colicky infants and 14 control infants at 2 months of age to estimate the function, balance and maturation of the autonomic nervous system [49]. The results showed no differences in the balance of the autonomic nervous system controlling heart rate variability between the colicky and control infants. The findings suggested that infantile colic is not due to a disturbance in the autonomic nervous system. Also, infants with colic have been shown to have a normal sleep pattern [50]. Barr suggested that colic might best be viewed as a clinical manifestation of normal emotional development, whereby an infant has not yet developed the capacity to regulate their crying episodes [51].

Parental Factors

A correlation has been noted between infantile colic and an increase in parental age as well as a high level of parental intelligence and education [9,39]. This may reflect a higher frequency of reporting or a lower level of tolerance of the symptoms of colic amongst such parents [4]. Several studies have shown an association between the incidence of infantile colic and psychological factors such as stressful pregnancies, postpartum depression, parental anxiety, dissatisfaction with the sexual relationship, and negative experiences during childbirth as well as poor parental skills [10,52,53].

Some investigators have found that infants who are exposed to maternal medications during labor, particularly epidural anesthesia and oxytocin, are more likely to develop colic than infants born following unmedicated labor [54,55]; while others disagree [10]. Further studies are necessary to confirm or refute these findings.

It has been suggested that colic may be the result of normal physiological events which trigger crying in infants who are temperamentally sensitive to these stimuli, and who then receive a parental response that is not appropriately soothing [15,18]. In one study, six infants whose parents were advised to let them cry showed no decrease in crying, whereas 20 infants whose parents were counseled on more effective responses showed a 70% decrease in crying [56]. These observations suggest that infantile colic may result at least in part from an inadvertent failure of the parents to respond to the infant's needs or wishes [15, 56].

Several studies have shown an increased risk of infantile colic when the mother smokes during the lactation period [32,57-59]. Matheson and Rivrud showed that breastfed infants of mothers who smoke have colic more frequently [57]. This can be explained by the presence of nicotine in breast milk and increased plasma and intestinal motilin level [41]. Reijneveld et al showed that maternal smoking increases the risk of infantile colic twofold but that breastfeeding weakens this association [32]. Said et al. found a correlation between infantile colic and parental smoking that was independent of the type of feeding [58]. Søndergaard et al observed a twofold increased risk of infantile colic among infants whose mothers smoked 15 or more cigarettes per day during their pregnancy or in the postpartum period [59]. Other investigators reported no association between colic and parental smoking but their results were based on relatively small study samples [7,60].

Clinical Manifestations

Characteristically, the onset of infantile colic is within two to three weeks of the expected due date (rather than the actual birth date) [1]. As such, the onset is usually delayed in premature infants [1,14,59]. The condition peaks when the child is at four to six weeks old.

Typically, infants with colic have prolonged bouts of crying and are inconsolable even by feeding. The infants are described as having clenched fists, flexed legs over their abdomen, an arching back, hard distended abdomen, flatulence, and a grimacing or a "pain" face that is usually flushed. These episodes usually occur late in the day or in the evening, and they may last for hours [4]. Not uncommonly, they may stop crying when they are picked up and cuddled. Frequent awakenings at night and extreme daytime wakefulness often occur in these infants [Rautava, 1995].

Clinical Evaluation

History

When documenting a medical profile during the clinical evaluation, the history should include a description of the timing of the onset of symptoms, the patterns of crying, the parents' response to the infant's crying, the intensity of crying, and any alleviating or aggravating factors that are associated with these episodes [14]. The infant's defecation, urination, and sleeping patterns should be evaluated for abnormalities. It would also be useful to have a description of the way in which the child is fed. If the infant is formula-fed, the content and preparation of the milk warrant attention. Information about the mother's diet is important if the infant is breast-fed. Any history of recurrent fever, vomiting, or trauma should be noted. The family history should include a description of the family structure, home environment, parental attitudes towards the infant, as well as family stresses and the support available to the family [15]. A prenatal history of drug exposure is important.

Physical Examination

Physical examination is important to exclude other possible causes of screaming and crying such as otitis media, urinary tract infection, fractured clavicle, corneal abrasion, foreign body in the eye, incarcerated hernia, malrotation of the bowel, intussusception, or anal fissure [15]. Weight, height, and head circumference should be plotted on standard growth charts since poor growth suggests the possibility of an underlying chronic systemic disorder. Vital signs should be noted. Fever indicates an underlying infection. Thus the examination should confirm the general well-being of the infant with infantile colic. Most parents are unconvinced by any attempt at reassurance that is not preceded by careful physical examination of their infant [15].

Differential Diagnosis

It is important to distinguish infantile colic from normal or physiological crying [56,62]. Brazelton observed the time spent crying in 80 healthy infants [48]. The average diary-recorded crying time was 1.75 hours/day in the 2nd week, 2.75 hours/day in the 6th week, and less than 1 hour/day by the 12th week [48]. Other conditions that may cause excessive crying include hunger, urinary tract infection, otitis media, gastroesophageal reflux, incarcerated hernia, malrotation of bowel, intussusception, fracture, corneal abrasion, foreign body in the eye, and anal fissure.

Investigations

Investigations are not required for the diagnosis of infantile colic. Appropriate laboratory tests and imaging studies may be indicated based on clinical findings if another cause is possible.

Complications

A crying infant often causes parents to ignore many of their own needs [63]. Colic may lead to parental stress, anxiety, fatigue, depression, anger, hostility, marital discord, guilt, feeling of helplessness, role ambivalence, and poor parent-child interaction [13,63]. At times, it may even lead to child abuse or domestic violence [13].

The effects of infantile colic on the family can be prolonged. In a study designed to assess the long-term effects of infants with colic on the family, Räihä et al found that families with infants who had severe colic, at one-year assessment, had more difficulties in communication, more unresolved conflicts, more dissatisfaction, and less empathy than families in the control group of noncolicky infants and infants with moderate colic [63]. However, when these children were seen again at 3 years of age, the family dynamics seemed to have normalized [64]. Rautava et al followed 338 colicky infants for 3 years [61]. Families of previously colicky infants were shown to be more dissatisfied with the arrangements of daily family responsibilities and with the amount of both leisure time and shared activities they were able to have. Children in the colic group had more sleeping problems and an increase in the frequency of temper tantrums than the control group. It was also found that fewer parents in the severe colic group than in the other groups had decided to have another child. Canivet et al. followed up on 50 formerly colicky infants and 102 controls at 4 years of age and evaluated their behavior, temperament, eating and sleeping habits, psychosomatic complaints, number of hospital admissions, growth, and family characteristics [65]. There were no significant differences between the two groups in most of the parameters studied. However, ex-colicky children displayed more negative emotions according to the temperament scale. They also had more negative moods during meals times with more complaints of stomach aches.

Savino et al conducted a prospective study on 103 infants between 31 and 87 days of age [66]. These infants were followed for 10 years. Ninety six infants completed the study (48 infants with severe colic and 48 infants without colic). Sleep disorders were found in 27 (56.3%) of the ex-colicky infants and in 6 (12.5%) of the subjects in the control group. Aggressiveness was found in 20 (41.7%) of ex-colicky infants and in 3 (6.2%) of subjects in the control group. Fussiness was found in 33 (68.7%) of ex-colicky infants and in 7 (14.6%) of the subjects in the control group. A feeling of supremacy was found in 18 (37.5%) of ex-colicky infants and in 2 (4.4%) of the subjects in the control group.

It is not clear whether colicky infants are more prone to have recurrent abdominal pain in childhood [67]. In the study by Savino et al., recurrent abdominal pain was observed in 16 (33.3%) of the ex-colicky infants and in 2 (4.4%) of the subjects in the control group [66]. Other authors did not notice the difference [68]. Future large-scale, well-designed studies will help to confirm or refute these findings.

In the same study by Savino et al, allergic rhinitis and allergic conjunctivitis were found in 13 (27.1%) of the ex-colicky infants and 2 (4.4%) of the subjects in the control group [66]. Pollenosis was found in 10 (20.1%) of the ex-colicky infants and in 2 (4.4%) of the subjects in the control group. Food allergy was observed in 11 (22.9%) of the ex-colicky infants and in 3 (6.2%) of the subjects in the control group. A Finnish study also suggested that infants with colic are more likely to develop atopic disorders compared with non-colicky infants [27]. In a prospective study of 983 infants, Castro-Rodriquez et al. did not find an increased risk of asthma or other atopic disorders in colicky infants [69].

Management

General Measures

The necessity of adequate feeding cannot be over-emphasized in the management of infantile colic. Feeding the infant in an upright position can prevent aerophagia. Care is to be taken when burping the infant, The hole in the bottle nipple should be the right size [4,13]. Holes or slits in the nipple should be made larger for the eager sucker; otherwise, the infant cannot get the milk fast enough and swallows excessive air. Bottles containing collapsible bags may decrease air swallowing [13].

It has been shown that healthy newborn infants often cry to obtain physical contact. Parents should be encouraged to pick up, cuddle, or carry their infant as much as they wish [4,14]. Increased sensory stimulation in the form of body massage, car-ride stimulation, and rocking or crib vibration has been found no more effective than placebo [70-73]. In fact, colicky infants may benefit from avoiding excessive stimulation [71,74].

Some parents become so exhausted by the crying that they need a break from the infant [4,14]. They should be advised to go out together and leave the infant with a babysitter. The parents should be encouraged to discuss their feelings and concern with each other to achieve mutual emotional support [15,20]. Although both parents may be in agony during the colicky crying sessions, they should be reassured by the physician of the benign nature of the condition and the fact that the infant will continue to thrive [15].

Dietary Manipulations

Breastfeeding mothers should continue breastfeeding [2,73]. Several randomized, controlled trials suggest a correlation between infantile colic in breastfed infants and their mothers' consumption of cow's milk and allergenic products [22,28,31]. Breastfeeding mothers with infants with severe colic or with atopic features should, with appropriate nutritional support, consider eliminating cow's milk from their diet and avoid potentially allergenic substances, such as caffeine, chocolate, eggs, and nuts [31,75,76].

Many randomized controlled trials have shown that hypoallergenic formulas may have a beneficial effect in the management of some formula-fed infants with infantile colic [5,6,21,24,77,78]. Most of the studies have, so far, involved a small sample size. Some of the studies have methodological flaws. It is hoped that future well-designed, large-scaled, randomized double-blind, placebo-controlled studies will provide more information in this area. A well-designed study should include the use of a common case definition, objective outcome measures, appropriate washout times in crossover trials, adequate blinding and repeated blind challenges of the proposed intervention to account for spontaneous resolution with increasing age [79,80]. Until results from such studies are available, no unequivocal recommendation can be made. The present consensus is that changing to a hypoallergenic formula is usually not necessary for formula-fed infants with mild to moderate colic [73]. Infants with severe colic, especially those with atopic features or a strong family history of atopy, may have a beneficial effect from hypoallergenic formulas such as whey or casein hydrolysates [2,75]. Completely hydrolysed hypoallergenic formulas are expensive and often unpalatable. Partially hydrolysed hypoallergenic formulas are less expensive and have a better taste [6,30]. However, partially hydrolysed hypoallergenic formulas contain slightly larger peptides and significantly more immunologically identifiable cow's milk thereby rendering them less desirable. Periodic challenges at monthly intervals are used to ensure that the improvement is related to dietary modification and not a result of natural resolution [2].

One small randomized controlled trial has found that soy formula reduced infant colic [81]. However, soy formula should be avoided as it has no proven value in the treatment of infantile colic [82-84] and some infants also react to soy with colic [85]. There is no evidence that low lactose milk formulas and fiber-enriched milk formulas are effective cow's milk substitutes [39,86].

Medications

Dicyclomine has been found to be an effective drug for the treatment of infantile colic as demonstrated in several double-blind studies [87,88]. However, this drug is now contraindicated in infants under 6 months of age due to reports of respiratory difficulties, apnea, seizures, coma, and death that have been associated with its use [14,15,87]. In a recent study, Savino et al found that cimetropium bromide was more effective than placebo in reducing the duration of crying in children with infantile colic [89]. These authors suggest that the use of anticholinergic drugs, aside from dicyclomine, should be re-evaluated for treatment of infantile colic. Cimetropium bromide is a quaternary ammonium semisynthetic derivative of the scopolamine. It acts as an antagonist of muscarine receptors of the visceral smooth muscles and it also has a direct myolytic activity [25].

Simethicone, a nonabsorbable defoaming agent with no systemic side effects, has also been used with success in some studies [90]. Simethicone reduces the surface tension of mucus, allowing entrapped air bubbles to coalesce and disperse. As a result, intestinal gas is expelled more easily. However, several randomized controlled studies found simethicone no more effective for the treatment of colic than the placebo [39,91,92].

Miscellaneous Treatments

It has been shown that some colicky infants have lower counts of intestinal lactobacilli [45,46]. It is postulated that *Lactobacillus reuteri*, a probiotic, helps to shift the intestinal ecological balance from potentially harmful flora to a beneficial one, thereby reducing the risk of gastrointestinal infections and allergic diseases [93,94]. Savino et al. randomly assigned 90 breastfed colicky infants to receive either *L. reuteri* (10^8 live bacteria per day) or simethicone (60 mg per day) for 28 days [36]. Eighty-three infants completed the study. Daily median crying times in the *L. reuteri* group and simethicone group were 159 minutes/day and 177 minutes/day, respectively, on the seventh day and 51 minutes/day and 145 minutes/day on the 28th day. On day 28, 39 (95%) patients in the *L. reuteri* group and 3 (7%) patients in the simethicone group were responders. No adverse events were reported. Although the preliminary results sound promising, the study sample size is too small to conclude. Until results of future well-designed, large-scaled, randomized, double-blind, placebo-controlled studies are available; no unequivocal recommendation can be made.

Some preliminary studies suggest that sucrose [95,96] and herbal teas [97,98] may alleviate colic. More data are required before the use of these agents can be recommended for the treatment of infantile colic. "Gripe water" is commonly used by parents to treat their colicky infants [72]. The product contains a variety of herbs and herbal oils such as cinnamon, cardamom, clove, ginger, peppermint, lemon balm, and licorice. However, there are no randomized controlled trials of "gripe water" in the treatment of infantile colic [73]. The use of herbal products is not entirely without risk and parents should be cautioned about their use in the treatment of infantile colic [25,97]. In a double-blind trial, Savino et al showed that a phytotherapeutic agent with *Matricariae recrutita*, *Foeniculum vulgare* and *Melissa officinalis* were effective in the treatment of infantile colic [99]. The authors suggested that the phytotherapeutic agent acts through its antispasmodic and antimeteoric activity.

There is insufficient evidence to justify chiropractic spinal manipulation in the treatment of infantile colic [73]. Wilberg et al. randomized 41 infants with infantile colic to receive two weeks of spinal manipulation versus two weeks of daily treatment with simethicone [100]. The mean reduction in crying from pre-treatment to days 8 to 11 was 2.7 hours and 1 hour in the spinal manipulation group and the simethicone group, respectively. The parents, however, were not blinded and this could be a potential source of bias. Olafsdottir et al randomized in a double-blind, placebo-controlled fashion 86 infants with infantile colic to chiropractic spinal manipulation or holding by a nurse [101]. Thirty two (69.9%) of 46 infants in the treatment group and 24 (60%) of 40 infants in the control group showed some degree of improvement. The authors conclude that chiropractic spinal manipulation is no more effective than placebo in the treatment of infantile colic.

Prognosis

In most infants, infantile colic resolves by 3 to 4 months of age, though colic has been shown to persist into the fourth or fifth month in up to 30% of cases [18].

Concluding Remarks

Infantile colic is a common condition that can be frustrating to both parents and physicians. The exact etiology is not known, although bovine milk proteins may play a role in many cases. Further research in this area is needed. Despite our ignorance, colicky infants continue to do well. If infantile colic has a purpose, it may be to teach us patience and humility.

REFERENCES

[1] Leung, AK; Lemay, JF. Infantile colic: a review. *J. R. Soc. Health*. 2004;124:162-166.

[2] Nutrition Committee, Canadian Paediatric Society. Dietary manipulations for infantile colic. *Paediatr. Child Health*. 2003:8:449-452.

[3] Wessel, MA; Cobb, JC; Jackson, EB; et al. Paroxysmal fussing in infants, sometimes called "colic". *Pediatrics*. 1954;14:431-434.

[4] Leung, AK. Infantile colic. *Am. Fam. Physician*. 1987;36:153-156.

[5] Jakobsson, I; Lothe, L; Ley, D; et al. Effectiveness of casein hydrolysate feedings in infants with colic. *Acta Paediatr*. 2000; 89:18-21.

[6] Lucassen, PL; Assendelft, WJ; Gubbels, JW; et al. Infantile colic: crying time reduction with a whey hydrolysate: a double-blind, randomized, placebo-controlled trial. *Pediatrics*. 2000;106:1349-1354.

[7] Canivet, C; Hagander, B; Jakobsson, I; et al. Infantile colic - less common than previously estimated? *Acta Paediatr*. 1996;85:454-458.

[8] Canivet, C; Jakobsson, I; Hagander, B. Colicky infants according to maternal reports in telephone interviews and diaries: a large Scandinavian study. *J. Dev. Behav. Pediatr*. 2002;23:1-8.

[9] Crowcroft, NS; Strachan, DP. The social origins of infantile colic: questionnaire study covering 76747 infants. *BMJ*. 1997;314:1325-1328.

[10] Høgdall, CK; Vestermark, V; Birch, M; et al. The significance of pregnancy, delivery and postpartum factors for the development of infantile colic. *J. Perinat. Med*. 1991;19:251-257.

[11] Lee, K. The crying pattern of Korean infants and related factors. *Dev. Med. Child. Neurol*. 1994;36:601-607.

[12] Reijneveld, SA; Brugman, E; Hirasing, RA. Excessive infant crying: the impact of varying definitions. *Pediatrics*. 2001;108:893-897.

[13] Lucassen, PL; Assendelft, WJ; van Eijk, TM; et al. Systemic review of the occurrence of infantile colic in the community. *Arch. Dis. Child*. 2001;84:398-403.

[14] Balon, AJ. Management of infantile colic. *Am. Fam. Physician*. 1997; 55:235-242.

[15] Leung, AK; Chan, PY; Cho, HY; et al. An updated review of infantile colic. *Can. J. Clin. Med*. 1997;4(10):16-19.

[16] Field, PA. A comparison of symptoms used by mothers and nurses to identify an infant with colic. *Int. J. Nurs. Stud*. 1994;31:201-215.

[17] St. James-Roberts, I; Halil, T . Infant crying patterns in the first year; normative and clinical findings. *J. Child. Psychol. Psychiatry*. 1991;32:951-968.

[18] Treem,WR. Infant colic: a pediatric gastroenterologist's perspective. *Pediatr. Clin. North Am*. 1994;41:1121-1138.

[19] Søndergaard, C; Skajaa, E; Henriksen, TB. Fetal growth and infantile colic. *Arch. Dis. Child*. 2000;83:F44-F47.

[20] Lehtonen, L; Korvenranta, H. Infantile colic: seasonal incidence and crying profiles. *Arch. Pediatr. Adolesc. Med*. 1995;149:533-536.

[21] Estep, DC; Kulczycki, A ,Jr. Treatment of infant colic with amino acid-based infant formula: a preliminary study. *Acta Paediatr*. 2000;89:22-27.

[22] Hill, DJ; Hudson, IL; Sheffield, LJ; et al. A low allergen diet is a significant intervention in infantile colic: results of a community-based study. *J. Allergy Clin. Immunol.* 1995;96:886-892.
[23] Iacono, G; Carroccio, A; Montalto, G; et al. Severe infantile colic and food intolerance: a long-term prospective study. *J. Pediatr. Gastroenterol. Nutr.* 1991;12:332-335.
[24] Lothe, L; Lindberg, T. Cow's milk whey protein elicits symptoms of infantile colic in colicky formula-fed infants: a double-blind crossover study. *Pediatrics.* 1989;83:262-266.
[25] Savino, F. Focus on infantile colic. *Acta Paediatr.* 2007;96:1259-1264.
[26] Hill, DJ; Hosking, CS. Infantile colic and food hypersensitivity. *J. Pediatr. Gastroenterol. Nutr.* 2000;30:S67-S76.
[27] Kalliomäki, M; Lappala, P; Korvenranta, H; et al. Extent of fussing and colic type crying preceding atopic disease. *Arch. Dis. Child.* 2001;84:349-350.
[28] Jakobsson, I; Lindberg, T. Cow's milk proteins cause infantile colic in breast-fed infants: a double-blind crossover study. *Pediatrics.* 1983;71:268-271.
[29] Hewson, P; Oberklaid, F; Menahem, S. Infant colic, distress, and crying. *Clin. Pediatr.* 1987;26:69-75.
[30] Leung, AK. Food allergy: a clinical approach. *Adv. Pediatr.* 1998;45:145-177.
[31] Hill, DJ; Roy, N; Heine, RG; et al. Effect of a low-allergen maternal diet on colic among breastfed infants: a randomized, controlled trial. *Pediatrics.* 2005;116:e709-e715.
[32] Reijneveld, SA; Brugman, E; Hirasing, RA. Infantile colic: maternal smoking as potential risk factor. *Arch. Dis. Child.* 2000;83:302-303.
[33] Miller, JJ; McVeagh, P; Fleet, GH; et al. Breath hydrogen excretion in infants with colic. *Arch. Dis. Child.* 1989;64:725-729.
[34] Moore, DJ; Robb, TA; Davidson, GP. Breath hydrogen response to milk containing lactose in colicky and noncolicky infants. *J. Pediatr.* 1988;113:979-984.
[35] Kanabar, D; Randhawa, M; Clayton, P. Improvement of symptoms in infant colic following reduction of lactose load with lactase. *J. Hum. Nutr.* 2001;14:359-363.
[36] Savino, F; Pelle, E; Palumeri, E; et al. *Lactobacillus reuteri* (American type culture collection strain 55730) versus simethicone in the treatment of infantile colic: a prospective randomized study. *Pediatrics.* 2007;119:e124-e130.
[37] Hyams, JS; Geertsma, MA; Etienne, NE; et al. Colonic hydrogen production in infants with colic. *J. Pediatr.*1989;115:592-594.
[38] Miller, JJ; McVeagh, P; Fleet, GH; et al. Effect of yeast lactase enzyme on "colic" infants fed human milk. *J. Pediatr.* 1990;117:261-263.
[39] Wade, S; Kilgour, T. Infantile colic. *BMJ.* 2001;323:437-440.
[40] Lothe, L; Lindberg, T; Jakobsson, I. Macromolecular absorption in infants with infantile colic. *Acta Paediatr. Scand.* 1990;79:417-421.
[41] Lothe, L; Ivarsson, SA; Ekmar, R; et al. Motilin and infantile colic: a prospective study. *Acta Paediatr. Scand.* 1990;79: 410-416.
[42] Shenassa, ED; Brown, MJ. Maternal smoking and infantile dysregulation: the case of colic. *Pediatrics.* 2004;114:e497-e505.
[43] Lehtonen, L; Svedström, E; Korvenranta, H . Gallbladder hypocontractility in infantile colic. *Acta Paediatr.*1994;83:1174-1177.

[44] Kurtoglu, S; Üzüm, K; Hallac, IK; et al. 5-Hydroxy-3-indole acetic acid levels in infantile colic: is serotoninergic tonus responsible for this problem? *Acta Paediatr.* 1997;86:764-765.

[45] Savino, F; Grassino, EC; Guidi, C; et al. Ghrelin and motilin concentration in colicky infants. *Acta Paediatr.* 2006;95:738-741.

[46] Savino, F; Cresi, F; Pautasso, S; et al. Intestinal microflora in breastfed colicky and non-colicky infants. *Acta Paediatr.* 2004;93:825-829.

[47] Savino F; Ballo, E; Oggero, R; et al. Bacterial counts of intestinal *Lactobacillus* species in infants with colic. *Pediatr. Allergy Immunol.* 2005;16:72-75.

[48] Brazelton, TB. Crying and colic. *Infant Ment. Health J.* 1990;11:349-356.

[49] Kirjavainen, J; Jahnukainen, T; Huhtala, V; et al. The balance of autonomic nervous system is normal in colicky infants. *Acta Paediatr.* 2001;90:250-254.

[50] Kirjavainen, J; Kirjavainen, T; Huhtala, V; et al. Infants with colic have a normal sleep structure at 2 and 7 months of age. *J. Pediatr.* 2001;138: 218-223.

[51] Barr, RG. Colic and crying syndromes in infants. Ped*i*atrics. 1998;102:1282-1286.

[52] Akman. I; Kuşçu, K; Özdemir, N; et al. Mothers' postpartum psychological adjustment and infantile colic. *Arch. Dis. Child.* 2006;91:417-419.

[53] Rautava, P; Helenius, H; Lehtonen, L. Psychosocial predisposing factors for infantile colic. *BMJ.* 1993;307:600-604.

[54] Murray, AD; Dolby, RM; Nation, RL; et al. Effects of epidural anaesthesia on newborns and their mothers. *Child Dev.* 1981;52:71-82.

[55] Thomas, DB. Aetiological associations in infantile colic: an hypothesis. *Aust. Paediatr. J.* 1981;17:292-295.

[56] Taubman, B. Clinical trial of the treatment of colic by modification of parent-infant interaction. *Pediatrics.* 1984;74:998-1003.

[57] Matheson, I; Rivrud, GN . The effect of smoking on lactation and infantile colic. *JAMA.* 1989;261:42-43.

[58] Said, G; Patois, E; Lellouch, J. Infantile colic and parental smoking. *BMJ.* 1984;289:660.

[59] Søndergaard, C; Henriksen, TB; Obel, C; et al. Smoking during pregnancy and infantile colic. *Pediatrics.* 2001;108:342-346.

[60] Haggart, M; Giblin, MJ Passive smoking and colic-like behaviour in babies. *Health Visitor.* 1998;61:81-82.

[61] Rautava, P: Lehtonen, L; Helenius, H; et al. Infantile colic: child and family three years later. *Pediatrics.* 1995;96:43-47.

[62] Hunziker, UA; Barr, RG. Increased carrying reduces infant crying: a randomized controlled trial. *Pediatrics.* 1986;77:641-648.

[63] Räihä, H; Lehtonen, L; Korhonen, T; et al. Family life 1 year after infantile colic. *Arch. Pediatr. Adolesc. Med.* 1996;150:1032-1036.

[64] Räihä, H; Lehtonen, L; Korhonen, T; et al. Family functioning 3 years after infantile colic. *J. Dev. Behav. Pediatr.* 1997;18:290-294.

[65] Canivet, C; Jakobsson, I; Hagander, B. Infantile colic. Follow-up at four years of age: still more "emotional". *Acta Paediatr.* 2000;89:13-17.

[66] Savino, F; Castagno, E; Bretto, R; et al. A prospective 10-year study on children who had severe infantile colic. *Acta Paediatr.* 2005;94(Suppl 449):120-132.

[67] Leung, AK; Lemay JF; Barker, CC. Recurrent abdominal pain in children. *Can. J. Diagn*. 2002;19(5):68-78.
[68] Joseph, AY; Lupu, GH. Recurrent abdominal pain and infantile colic. *Am. J. Dis. Child.* 1984;138:990-991.
[69] Castro-Rodriquez, JA; Stern, DA; Halonen, M; et al. Relation between infantile colic and asthma atopy: a prospective study in an unselected population. *Pediatrics*. 2001;108:878-882.
[70] Barr, RG; McMullan, SJ; Spiess, H; et al. Carrying as colic "therapy": a randomized controlled trial. *Pediatrics*. 1991;87:623-630.
[71] Huhtala, V; Lehtonen, L; Heinonen, R; et al. Infant massage compared with crib vibrator in the treatment of colicky infants. Pediatrics. 2000;105,e84.
[72] Roberts, DM; Ostapchuk, M; O'Brien, JG. Infantile colic. *Am. Fam. Physician*. 2004;70:735-742.
[73] Wade, S. Infantile colic. *Clin. Evid.* 2006;15:439-447.
[74] McKenzie, S. Troublesome crying in infants: effect of advice to reduce stimulation. *Arch. Dis. Child*. 1991;66,1416-1420.
[75] Sampson, HA; Sicherer, SH; Bimbaum, AH; et al. AGA technical review on the evaluation of food allergy in gastrointestinal disorders. *Gastroenterology*. 2001;120:1026-1040.
[76] Schach, B; Haight, M. Colic and food allergy in the breastfed infants: is it possible for an exclusively breastfed infants to suffer from food allergy? *J. Hum. Lact.* 2002;18:50-52.
[77] Forsyth, BW. Colic and the effect of changing formulas: a double-blind, multiple-crossover study. *J. Pediatr*. 1989;115:521-526.
[78] Savino, F; Palumeri, E; Castagno, E; et al. Reduction of crying episodes owing to infantile colic: a randomized controlled study on the efficacy of a new infant formula. *Eur. J. Clin. Nutr*. 2006;60:1304-1310.
[79] Garrison, MM; Christakis, DA. Early childhood: colic, child development, and poisoning prevention. A systemic review of treatments for infant colic. *Pediatrics*. 2000;106:184-190.
[80] Sampson, HA. Infantile colic and food allergy: fact or friction? *J. Pediatr*. 1989;115:583-584.
[81] Campbell, JP. Dietary treatment of infantile colic: a double-blind study. *J. R. Coll. Gen. Pract.* 1989;39:11-14.
[82] Agostoni, C; Axelsson, I; Goulet, O; et al. Soy protein infant formulae and follow-on formulae: a commentary by the ESPGHAN Committee on Nutrition. *J. Pediatr. Gastroenterol. Nutr*. 2006;42:352-361.
[83] Committee on Nutrition, the American Academy of Pediatrics. Soy protein-based formulas: recommendations for use in infant feeding. *Pediatrics*. 1998;101:148-153.
[84] Turck, D. Soy protein for infant feeding: what do we know? *Curr. Opin. Clin. Nutr. Metabol. Care.* 2007;10:360-365.
[85] Lindberg, T. Infantile colic and small intestine function: a nutritional problem? *Acta Paediatr*. 1999;88:58-60.
[86] Lucassen, PL; Assendelft, WJ; Gubbels, JW; et al. Effectiveness of treatments for infantile colic: systemic review. BMJ. 1998;316:1563-1569.

[87] Parkin, PC; Schwartz, CJ; Manuel, BE. Randomized controlled trial of three interventions in the management of persistent crying of infancy. *Pediatrics*. 1993;92:197-201.

[88] Weissbluth, M; Christoffel, KK; Davis, AT. Treatment of infantile colic with dicyclomine hydrochloride. *J. Pediatr.* 1984;104:951-955.

[89] Savino, F; Brondello, C; Cresi, F; et al. Cimetropium bromide in the treatment of crisis in infantile colic. *J. Pediatr. Gastroenterol. Nutr.* 2002;34:417-419.

[90] Sethi, KS; Sethi, JK. Simethicone in the management of infant colic. *Practitioner*. 1988;232:508.

[91] Danielsson, B; Hwang, CP. Treatment of infantile colic with surface active substance (simethicone). *Acta. Paediatr. Scand.* 1985;74:446-450.

[92] Metcalf, TJ; Irons, TG; Sher, LD; et al. Simethicone in the treatment of infantile colic: a randomized, placebo-controlled, multicenter trial. *Pediatrics*. 1994;94:29-34.

[93] Neu, J. Probiotics: protecting the intestinal ecosystem? *J. Pediatr.* 2005;147:143-146.

[94] Weizman, Z; Asli, G; Alsheikh, A. Effects of a probiotic infant formula on infections in child care centers: comparison of two probiotic agents. *Pediatrics*. 2005;115:5-9.

[95] Akçam, M; Yilmaz, A. Oral hypertonic glucose solution in the treatment of infantile colic. *Pediatr. Int.* 2006;48:125-127.

[96] Markestad, T. Use of sucrose as a treatment for infant colic. *Arch. Dis. Child.* 1997;76:356-358.

[97] Crotteau, CA; Wright, ST. What is the best treatment for infants with colic? *J. Fam. Pract.* 2006;55:634-636.

[98] Weizman, Z; Alkrinawi, S; Goldfarb, D; et al. Efficacy of herbal tea preparation in infantile colic. *J. Pediatr.* 1993;122:650-652.

[99] Savino, F; Cresi, F; Castagno, E; et al. A randomized double-blind placebo controlled trial of a standardized extract of *Matricariae recrutita*, *Foeniculum vulgare* and *Melissa officinalis* (ColiMil®) in the treatment of breastfed colicky infants. *Phytother. Res.* 2005;19:335-340.

[100] Wilberg, JM; Nordsteen, J; Nilsson, N. The short-term effect of spinal manipulation in the treatment of infantile colic: a randomized controlled clinical trial with a blinded observer. *J. Manipulative Physiol. Ther.* 1999;22:517-522.

[101] Olafsdottir, E; Forshei, S; Fluge, G; et al. Randomized controlled trial of infantile colic treated with chiropractic spinal manipulation. *Arch. Dis. Child.* 2001;84:138-141.

In: Food Allergies: New Research
Editor: Carrie M. Chesterton, pp. 173-187
ISBN 978-1-60456-978-0

Chapter 6

EOSINOPHILIC GASTROINTESTINAL DISORDERS

***Thomas P. Miller*[1] *and Alexander K.C. Leung*[2,*]**

[1]Michigan State University College of Human Medicine, Allergy Associates of Western Michigan P.C., 360 East Beltline, NE Ste 100, Grand Rapids, MI, 49509-1208 USA

[2]Department of Pediatrics, University of Calgary, Alberta Children's Hospital, #200, 233 – 16th Avenue NW, Calgary, Alberta, T2M 0H5, Canada

ABSTRACT

Eosinophilic gastrointestinal disorders consist of diseases involving eosinophilic infiltration of the gastrointestinal tract. These include eosinophilic esophagitis (EE), eosinophilic gastroenteritis, and eosinophilic colitis. Much of the recent literature has described eosinophilic esophagitis with a relative lack of information regarding eosinophilic gastroenteritis and eosinophilic colitis specifically. Many studies of EE reviewed in this chapter included patients with extra-esophageal eosinophilic involvement. Much of what is known regarding EE is applicable to eosinophilic gastroenteritis, and perhaps to a lesser extent eosinophilic colitis. This chapter focuses mostly on eosinophilic esophagitis; however, the concepts are applicable to all the eosinophilic gastrointestinal disorders. Though first described decades ago, these entities are still, to some degree, poorly defined. As such, the diagnosis is described as clinicopathologic in that it relies on clinical signs and symptoms with supporting laboratory and histologic findings. As background, this chapter includes a brief description of the gastrointestinal tract barrier and oral tolerance. The gastrointestinal tract serves as a physical and immunological barrier. The normal response of oral tolerance reflects a lack of immunologic responsiveness as a result of prior exposure. Food hypersensitivity responses occur after a failure of oral tolerance development. This is important for all types of food hypersensitivity responses including the eosinophilic gastrointestinal disorders. Unlike immediate hypersensitivity (type I IgE-mediated) food allergic responses, eosinophilic esophagitis, eosinophilic gastroenteritis, and eosinophilic colitis may involve both IgE- and non-IgE-mediated responses. This is also in contrast to food protein-induced enterocolitis/colitis or celiac disease in which a cell-mediated mechanism is likely responsible without evidence of IgE involvement. In addition to the

[*] Correspondence to: Dr. Alexander K.C. Leung, #200, 233 – 16th Avenue NW, Calgary, Alberta, Canada T2M 0H5; Telefax: (403) 230-3322, e-mail:aleung@ucalgary.ca

background information, this chapter describes the pathogenesis, clinical manifestations, as well as therapeutic approach to the eosinophilic gastrointestinal disorders.

Keywords: eosinophilic gastrointestinal disorders, eosinophilic esophagitis

INTRODUCTION

The eosinophilic gastrointestinal disorders (EGID) are increasingly recognized diseases characterized by eosinophilic infiltration of the gastrointestinal tract in the absence of an underlying cause. The development of this condition is likely the result of a complex interplay between genetic predisposition, environmental exposure (especially food antigens), the gastrointestinal (immunologic) barrier, failure of oral tolerance development, and finally immunologic response characterized by a Th2 profile and eosinophilic infiltration. These relationships are beginning to be understood in the setting of eosinophilic esophagitis (EE) and likely apply to varying degrees to the other eosinophilic gastrointestinal disorders.

GASTROINTESTINAL BARRIER

The gastrointestinal tract serves as a physical barrier operating at the host-environment interface. Of greater importance is the role that it plays as a site of mucosal immune interaction. This mucosal immune system is also described as the gut-associated lymphoid tissue (GALT) system. It is imperative for normal health that the intestinal immune system is able to mount a protective immune response to pathogens while maintaining a state of tolerance (non-responsiveness) to normal bacterial flora and food antigens. A brief review of this immune barrier has recently been published [1]. We can consider the immune response as consisting of inductive and effector phases. The Peyer's patches and mesenteric lymph nodes are the main locations of the cellular components of the intestinal immune system. These are likely the main locations for induction of immunologic reactivity or tolerance. The Peyer's patches are lymphoid aggregates consisting of large B cell follicles and interfollicular T cell regions, as well as numerous dendritic cells and macrophages. These areas are found beneath a single layer of columnar epithelial cells referred to as follicle associated epithelium (FAE). Microfold (M) cells are found in this FAE (as well as other areas lacking these lymphoid aggregates) [2]. These M cells function to directly transport antigens from the lumen to the subepithelial lymphoid tissue. These cells have been shown to transport bacteria, viruses, proteins, and noninfectious particles [3, 4]. Dendritic cells can function in conjunction with the M cells, acquiring antigen at their basolateral surface. There is some evidence that dendritic cells may also be able to sample antigen directly from the luminal surface [5]. The dendritic cells are able to migrate within the Peyer's patches as well as distant sites such as mesenteric lymph nodes for antigen presentation. Thus dendritic cells likely play an important role in the induction of protective immune responses as well as oral tolerance [6]. Dendritic cells may also migrate to the lamina propria of the intestine to facilitate an immune response. Other elements of the effector component of the immune response include mast cells, eosinophils and lymphocytes. Mast cells are found throughout the gastrointestinal tract and are important in host response to parasitic infections and likely also are involved in the immune response to bacteria that compromise the epithelial barrier [7]. Eosinophils are normally found in the lamina propria of the stomach as well as small and large intestines; they are not normally found in the esophagus.

Eosinophil Function

In addition to the pro-inflammatory effect of eosinophils, it appears that eosinophils may play a significant role in the normal functioning of the intestinal tract. Eosinophils have been recognized as playing a role in host defense to parasitic infections. In addition to this better known role it is likely that eosinophils interact with and regulate T cell lymphocytes. Eosinophils may also have a role in the normal gestational development of the gastrointestinal tract. These topics are discussed in greater detail in a recent review of eosinophilic gastrointestinal disorders [8].

Oral Tolerance

Oral tolerance has recently been reviewed [9]. In summary, tolerance of dietary protein is dependent upon protein modification in the lumen, processing by antigen presenting cells, interaction with regulatory T cells and resulting suppression of the immune response. Digestive processing of proteins involves proteases from the stomach, pancreas, and small intestine. These proteases break down most dietary proteins to free amino acids, di- or tripeptides. The action of digestive enzymes has been shown to increase the oral tolerance of dietary proteins [10]. Intestinal epithelial cells can also play a role in digesting soluble dietary protein after endocytosis and phagolysosome formation. Dendritic cells, acquiring antigen from the M cells or directly from the lumen, transport antigen for presentation to B cells in Peyer's patches. Transforming growth factor-ß (TGF-ß) secreting T cells mediate IgA switching of these B cells. This promotes oral tolerance [11-15]. Oral tolerance can be the result of T cell anergy or deletion after high dose exposure or suppression of the immune response by regulatory T cell cytokines (TGF- ß or IL-10) after low dose exposure [9]. It is also likely that genetic influences play a role as well as the bacterial flora of the intestines.

Clinical Features

Perhaps the most common presenting symptoms in individuals with eosinophilic esophagitis are dysphagia and symptoms of gastroesophageal reflux disease (GERD) such as recurrent regurgitation and heartburn. A recent systematic review in adults [16] noted the most common presenting symptoms included dysphagia (93%), food impaction (62%), and heartburn (23%). The age of the patient will impact the specific manifestations of this. For example, feeding refusal or 'irritability' with food may occur in children too young to describe dysphagia. Likewise GERD symptoms of heartburn or water brash may be ill-defined in young children. In older children and adults these symptoms are more easily described. Emesis and abdominal pain have been reported in both children and adults, though this seems to be more common in children. Food impaction has been reported in both age groups, though this is more common in adults. Other signs or symptoms can include failure to thrive in children or chest pain, diarrhea, or weight loss in either age group. EE should be considered whenever these symptoms are present, but especially in the setting of 'refractory' GERD, as the symptoms of EE will be, at best, partially responsive to anti-reflux therapy. Gross endoscopic findings of EE frequently include linear furrowing (or vertical lines visible in the esophageal mucosa), white exudates (which could also appear as specks or nodules), and/or circular rings (transient or fixed) [17].

Patients with EGID having extra-esophageal involvement may present with various symptoms or signs. These can include abdominal pain or irritability, dysphagia, emesis, food impaction, gastric dysmotility, anemia, diarrhea, or hypoproteinemia. As eosinophilic inflammation can involve the entire gastrointestinal tract, the symptoms can be varied and non-specific. Eosinophilic gastroenteritis is distinct from EE in that the eosinophilic infiltrate can involve the mucosal, muscularis, or serosal layers [18], therefore biopsies may be normal if the eosinophilic infiltrate is restricted to the muscularis or serosa. Eosinophilic infiltration of the muscularis may lead to symptoms of gastric outlet obstruction similar to pyloric stenosis or small bowel obstructive symptoms from thickening of the intestinal wall. Exudative ascites has been observed with prominent serosal infiltration [19]. Food allergy or intolerance may be present in up to 50% of patients with the mucosal form of eosinophilic gastroenteritis. It is likely that eosinophilic colitis is a non-IgE mediated disease in contract to eosinophilic esophagitis or eosinophilic gastroenteritis. The exact immunologic mechanism is unknown, though evidence points to a T cell process [20].

DEFINITION

A recent systematic review and consensus recommendation from the American Gastroenterological Association (AGA) Institute [17] defined eosinophilic esophagitis (for the purpose of that review) as a primary clinicopathologic disorder of the esophagus, characterized by symptoms (esophageal and/or upper gastrointestinal) associated with $\geq$ 15 eosinophils/high power field (HPF) in esophageal mucosal biopsy specimens. Some authors have used a cut-off of $\geq$ 20-24 eosinophils/HPF [21-24]. GERD must be ruled out by either a normal 24 hour pH probe or a lack of response to high dose proton pump inhibition (up to 2 mg/kg/day).

EPIDEMIOLOGY

The AGA review summarized 13 studies describing the epidemiology of EE in adults and 16 studies for children. They found that of the 323 adult patients, most were male (76%) with a mean age of 38 years (range 14 to 89). In children (n = 754) the gender distribution was similar (66% male) with a mean age of 8.6 (range 0.5 to 21.1). A different systematic review [16] evaluating EE in adults only found a male predominance (male:female ratio 3:1) in 325 patients from 24 studies. EE has been described in individuals of many ethnic backgrounds, however, there is little information regarding comparative prevalence for ethnic groups. EE is reported most commonly in developed countries, which may simply reflect selection, however as EE has an association with atopy it is interesting to postulate that the true prevalence of EE might be higher in developed countries similar to the higher prevalence of atopy.

The prevalence of EE described in the literature is variable. The prevalence of EE has been noted to range from 8.9/100,000 population to 9.3% in children and 30/100,000 to approximately 50% in adults. The lower numbers typically reflect prevalence data at referral centers serving as the sole providers for a geographic region with the denominator comprising the regional population. The higher numbers typically reflect prevalence in a highly selected population. For example Liacouras et al. [25] diagnosed EE in 9.3% of children with eosinophils noted on esophageal biopsy, and 1.1% of the population that presented to their tertiary referral center with reflux symptoms. Kerlin et al. [26] found eosinophilic esophagitis in 48% of adult patients undergoing endoscopy for food impaction. Desai et al. found eosinophilic esophogitis in 54% of a similar population of patients [27]. The true prevalence of eosinophilic esophagitis is unknown, however, there was a recent study describing upper endoscopy findings in a random sample of Swedish adults [28]. In this study, Ronkainen et al. performed upper endoscopy with biopsies of the distal esophagus on 1,000 individuals. These individuals were from 2,122 respondents to a mailed questionnaire (Abdominal Symptom Questionnaire) that had been sent to 2,860 adults. The 1,000 individuals who were willing to undergo endoscopy were similar in age and sex distribution as the 2,122 respondents, though it is unclear whether questionnaire results were similar between those willing to undergo endoscopy compared to those that were not. Of the 1,000 individuals biopsied, 48 were found to have eosinophils and 4 were found to have enough eosinophils present to be categorized as having definite eosinophilic esophagitis. The authors defined definite eosinophilic esophagitis as a biopsy specimen containing $\geq$ 20 eosinophils/HPF. It is unknown whether individuals who responded to the survey were more apt to have abdominal symptoms. It is unknown whether those willing to undergo endoscopy were more likely to be symptomatic. It is also unknown whether the individuals defined as having definite EE had GERD as 24-hour pH probe information was not available. However, the results of this study suggest a prevalence of EE of approximately 0.4% in western Caucasian adults.

A major contributor to the expanding knowledge base of EE has been the group at the Cincinnati Children's Hospital. A recent update from their pediatric patient population (0 to19 years) estimated a prevalence of approximately 0.09% [29]. This represents their cases out of their service population, as they are the only pediatric gastroenterology center in their region. They also defined EE as having $\geq$ 24 eosinophils/HPF. In their series, the investigators followed EE prevalence in one county in which they had data from 2000 to 2006. Over this seven-year period of time the prevalence for that county in southern Ohio increased from 0.991/10,000 to the present 9.076/10,000 (0.09%). This is a follow-up of this population revealing a continued increase from the previous observation period of 2000 to 2004 where the observed increase was 0.991/10,000 to 3.106/10,000 [21]. Other investigators using similar methodologies (tracking the number of cases determined at a regional referral center divided by the regional population) have found similar results. Cherian et al [30] observed an increase in prevalence in Western Australia from 0.05/10,000 in 1995 to 0.89/10,000 in 2004.

There is conflicting evidence as to whether EE is actually increasing in incidence or is simply being recognized more frequently. A recent study by Vanderheyden et al [31] suggested that the increasing incidence of EE is likely related to increased recognition and not an absolute increase in the underlying prevalence of disease. Consecutive patients who underwent esophageal mucosal biopsies from May through June during 2005 were compared to the same time period during 1990. Patients with Barrett metaplasia or carcinoma were excluded. Assuming that the prevalence of EE was increasing while other esophageal conditions were not, the relative percent of EE cases should be higher in the 2005 sample compared with 1990. This was not the case as the percent EE as defined by $\geq$ 20 eosinophils/HPF was 7% in each group (8 of 115 for 1990, 10 of 150 for 2005). Of the 18 EE cases, there was a male predominance (male:female 3.5:1) which is similar to other studies. Another recent study revealed similar results. Franciose et al [32] determined the proportion of children with severe EE over the past 20 years utilizing current criteria. The authors reviewed upper endoscopy biopsies from 1985, 1995, and 2005. Isolated EE as defined by $\geq$ 15 eosinophils/HPF was found in 3 of 86 (3.5%) in 1985, 33 of 476 (6.9%) in 1995, and 83 of 1273 (6.5%, $p > 0.2$) in 2005. The number of endoscopies with multiple biopsies was also lower in 1985. These authors concluded that there seems to be no significant increase in the proportion of children with EE over the last 20 years.

Pathogenesis

The etiology of EE is not clear. At the molecular level it is clear that EE is a Th2 dependent process [33]. Th2 refers to a subset of CD4+ T helper (thus Th) lymphocytes. The Th2 cells are involved in humoral responses (vs cell mediated responses for Th1 cells), typically to allergens or parasites (vs microbes for Th1 cells). The major cytokines secreted by Th2 cells include IL-4, IL-5, IL-6, IL-10, and IL-13 (vs IFN-gamma, IL-2 from Th1). IL-4 and IL-13 are the main cytokines that promote IgE isotype switching (which is inhibited by IFN-gamma). IL-5 attracts, activates, and prolongs survival of eosinophils. Recent investigations have demonstrated the importance of IL-5 in the induction and remodeling in EE [34]. IL-13 has also been implicated in driving the pathogenesis and remodeling observed in EE [35]. Eotaxin-3, which is a chemokine induced by IL-13, has been found to be upregulated in EE patients [36]. Besides the increased eosinophilic infiltration, there may be other cellular abnormalities occurring as well. A recent report [37] suggested that under the conditions of eosinophilic esophagitis, esophageal epithelial cells might be capable of antigen presentation, which would be a novel immunologic role. It has recently been observed, on distal esophageal biopsy specimens, that basophils are also found in increased numbers in pediatric EE patients compared to patients with GERD or normal controls [38]. Their role in the pathogenesis of EE is unknown.

There is a clear association between EE and allergies. It is interesting that most patients with EE show evidence of food allergies (positive skin test or RAST results). However, few of these patients have a history of anaphylaxis to the implicated foods leading some experts to postulate that EE involves a mechanism distinct from the classic IgE-mediated immediate hypersensitivity response [39]. However, this characteristic (frequent food sensitivities infrequently manifested by anaphylaxis) is similar to atopic dermatitis. The role of food allergies in EE was fairly well established by Kelly et al. in 1995 [40]. These authors studied a series of 10 children with refractory GERD. Affected children continued to experience symptoms despite antacids and fundoplication. Biopsies revealed persistent eosinophilic infiltration. The children were given an amino-acid based formula (plus corn and apples in those old enough) for six weeks. This improved symptoms (80% complete resolution) and histology (median eosinophils/HPF decreased from 41 to 0.5). A subsequent open food challenge resulted in return of the same symptoms that the children had been previously experiencing. A causative link between food allergies and EE has subsequently been confirmed by other investigators [41, 42]. As mentioned above, some investigators postulate both IgE-mediated and non-IgE-mediated mechanisms are involved in the pathogenesis. The study by Kelly et al. [40] demonstrated that specific foods were capable of prompting symptoms despite negative skin prick tests (SPT). This may reflect non-IgE-mediated mechanisms or an issue of sensitivity for skin prick tests in this setting. For aeroallergen testing, negative skin prick testing is typically followed by intradermal testing to increase sensitivity. At present intradermal testing is not recommended for food allergy testing.

Atopy patch testing (APT) has been suggested to increase the sensitivity of SPT alone [43, 44]. APT has been used commonly to diagnose contact sensitivities that are thought to involve non-IgE-mediated, cell-mediated delayed type hypersensitivity responses. The studies that have evaluated APT in addition to skin prick testing have suggested that a few foods, not identified by skin testing, can be identified by APT. Two recent studies in children with EE revealed that APT identified at least 1 additional food not identified by SPT in most children [45, 46]. There has also been concern raised that skin testing with commercial food extract may be less sensitive than testing with fresh foods [47]. It certainly may be that the additional foods detected by APT reflect a non-IgE mechanism. However, this additional sensitivity of APT may in fact be due to increased sensitivity as most centers use fresh food for APT. These investigators have not involved skin biopsies of the patch test site, which might help clarify the immune mechanism involved.

The association between allergic disease and EE which has been observed in children appears less strong in adults. A small study evaluating the role of APT in pediatric and adult patients with EE and eosinophilic colitis found adults to be less apt to have a positive APT than the children [48]. Two recent studies in adults [49, 50] found less dramatic improvement with avoidance of implicated foods compared with the pediatric trials. Our inability to establish this causal relationship between foods and EE in all patients likely relates to a lack of sensitivity of our testing methods, our lack of understanding of all the mechanisms involved, or both.

There is also an association between GERD and esophageal eosinophilic infiltration. It appears that GERD alone leads to eosinophilic infiltration of the mucosa, though in general the numbers of eosinophils is lower in GERD than in EE. It could be that this eosinophilic infiltration is in response to the gastric acid exposure associated with GERD. However, in addition to the gastric acid, the esophagus is exposed to stomach contents including food. Thus it is possible that this eosinophilic infiltration of GERD could be related to food sensitivity as well.

There is also an association between aeroallergen sensitization and EE. This was initially postulated in response to animal studies where it was observed that intranasal exposure to allergen resulted in esophageal eosinophilic infiltration [51]. A case report of a patient with EE who had seasonal exacerbations of EE during the pollen allergy season despite being on a constant diet suggested that this relationship may exist in humans as well [52]. A recent report evaluating season of first biopsy and geographic distribution (as a surrogate of aeroallergen exposure) found no difference between EE patients and patients requiring endoscopy and biopsy who were not found to have EE, thus suggesting that, in that population, aeroallergens were likely not playing a role in EE [53]. It is likely that aeroallergens do not play a major role in EE as many of the children who had a dramatic improvement in disease with dietary manipulation also had inhalant allergies that were not treated.

Natural History

EE is a chronic relapsing disease. A recent review estimated the recurrence rate between 25 and 40% after successful induction of remission [57]. Assa'ad et al. [58] recently published an 8-year follow-up study of 89 pediatric patients with EE. Most of these patients also had eosinophilic involvement beyond the esophagus. Eosinophilic involvement of the stomach and duodenum was noted in 47% and 57%of the cases, respectively. Involvement of the colon was noted in 63% on those undergoing colonoscopy (n = 38). In this patient population, symptom resolution occurred in 66%. Of these patients 79% had a recurrence over a mean of 3 years. In a larger pediatric EE population (n = 381), Liacouras et al. [56] describe their experience treating EE patients over a 10-year period of time. After withdrawal of medical therapy, most patients experienced a return of their symptoms and esophageal eosinophilia. Three studies in adults [59-61] confirm that EE appears to be a chronic condition in adults as it is in children. The primary initial therapy in these studies was stricture dilation in contrast to other modes of therapy in the pediatric studies. After initial improvement, many patients experienced a recurrence of symptoms (most frequently dysphagia) and either underwent repeat stricture dilations or medical therapy (most commonly topical corticosteroids).

Differential Diagnosis

Eosinophilic disorders of the gastrointestinal tract can be divided into primary and secondary disorders. The primary EGID are characterized by increased eosinophilic infiltration of the gastrointestinal mucosa without any known cause such as parasitic infection, drug reaction, malignancy, inflammatory bowel disease, or hypereosinophilic syndrome. There are also no pathognomonic findings or blood tests that diagnose EGID. The evaluation of abdominal complaints should start with a complete history and physical examination. If no obvious causes are identified, endoscopic evaluation will frequently identify intraepithelial eosinophilic infiltration. At this point, ruling out secondary causes of eosinophilic inflammation is required before the diagnosis of EGID is confirmed. This includes evaluation for drug hypersensitivity, malignancy, collagen-vascular disease, or infection. As eosinophils can be found normally in the gastrointestinal tract (except the esophagus), experts have suggested several factors to consider when differentiating EGID from normal conditions [8, 54]. These include: (1) eosinophil quantification (compared to institution specific normal values), (2) abnormal eosinophil location (such as intraepithelial, superficial mucosal, or intestinal crypt region), (3) associated pathologic abnormalities (such as epithelial hyperplasia found with EE) (4) absence of other pathologic findings suggestive of other disorders. Eosinophils can normally be found in the stomach and intestine, therefore the diagnosis of EGID is more complex than in EE. It is also recognized that many disorders can be associated with eosinophilic infiltration of the stomach including infection with *Helicobacter pylori* or other bacteria or parasites, inflammatory bowel disease, collagen-vascular disease including vasculitis or scleroderma, myeloproliferative disorders, hypereosinophilic syndrome, or drug hypersensitivity [55]. As with eosinophilic gastroenteritis, eosinophilic colitis can result from parasitic infections, drug reactions, inflammatory bowel disease, vasculitis etc. Protein-induced (or allergic) proctocolitis is a common cause of bloody stools in infancy. Milk and soy proteins are the most frequently implicated causes. When EGID is confirmed, the location of the abnormal eosinophilic inflammation will determine whether the patient has EE, eosinophilic gastritis/gastroenteritis, or eosinophilic colitis.

Diagnostic Approach

The diagnostic approach and treatment for EE outlined here is summarized from the review and consensus recommendations [17] mentioned above. Endoscopic evaluation is required. If a diagnosis of EE is being considered, upper endoscopy with biopsies should be undertaken. Gross appearance should be documented and photographed. Biopsies should be obtained along the length of the esophagus, as well as from the stomach and duodenum. Normal appearing mucosa should not discourage this practice as in one series the esophageal mucosa appeared normal in 30% of children with EE [56].

Monitoring pH may or may not be required. If it is not apparent whether EE or GERD is the primary diagnosis after endoscopy, pH monitoring may be considered. Repeat endoscopy after 6 to 8 weeks of therapy consisting of high dose proton pump inhibition may be considered as an alternative as the eosinophilia associated with GERD should be significantly improved.

Radiography may be considered. An upper GI contrast study may be appropriate in patients presenting with dysphagia or vomiting to identify stricture or malrotation. This information can be helpful before upper endoscopy, however, this is unnecessary in patients presenting with typical GERD symptoms.

A complete allergy evaluation should be obtained by an allergist to determine the presence of food sensitivities as well as associated allergic conditions including inhalant sensitivities, allergic rhinitis, etc. At present, standard allergy testing consists of SPT. Some investigators supplement the SPT with APT for foods found negative on SPT. The addition of APT is not practiced in all centers, though this may be the case in the future.

Treatment

Dietary manipulation should be aggressively pursued. In general, dietary manipulation can take one of three forms. The first option consists of simply avoiding six of the most common allergenic foods (cow-milk protein, soy, wheat, egg, peanut, and seafood) [42]. The second option involves avoiding foods implicated by allergy testing. The third option involves utilizing an elemental formula. These various options should be tailored to the clinical situation. For example, if skin testing does not result in the identification of many foods, it might not be unreasonable to pursue a trial avoiding the six allergenic foods. The studies that revealed benefit of the amino acid-based formula typically utilized nasogastric or gastrostomy tubes because of the poor taste of the formula. Hydrolyzed formula is not adequate. Dietary manipulation has not been shown to be as effective in adults. Dietary manipulation should be done with the consultation of a registered dietician to ensure adequate caloric and nutritional intake is maintained.

Steroids are frequently necessary to induce and maintain remissions. Both systemic and topical corticosteroids have demonstrated efficacy in improving symptoms as well as histology in patients with EE. Due to potential side effects, systemic corticosteroids should be given on a temporary basis for urgent situations such as severe dysphagia requiring hospitalization or when there is evidence of dehydration or weight loss. Chronic steroid therapy should be administered topically in the form of a metered dose inhaler (MDI) without a spacer. The MDI should be actuated using the closed mouth method (inserted in the mouth with lips sealed around the mouthpiece). Instead of inhaling, however, the medication should be swallowed without rinsing afterward (no eating or drinking for 30 minutes). The consensus recommendations [17] for dosage are 440-880 μg/day in children and 880-1760 μg/day in adolescents or adults in two to four divided doses. Patients should continue for at least 6 to 8 weeks. Cromolyn sodium and leukotriene receptor antagonists are not recommended.

Comorbid conditions should be treated. Treatment of GERD may be useful in establishing the diagnosis of EE and it may partially improve symptoms. Symptomatic improvement may also occur with stricture dilation. Treating comorbid conditions should be considered for symptomatic benefit, recognizing that this will not be a primary treatment for EE.

The treatment of eosinophilic gastritis/gastroenteritis or eosinophilic colitis are similar to EE except that drug delivery strategies must be targeted to the site of intestinal involvement and IgE related triggers are rarely the cause of colitis. Eosinophilic colitis is frequently a secondary condition and treatment of the underlying condition is essential. Eosinophilic colitis of infancy is typically a benign condition that resolves with removal of the protein trigger. Steroids or non-steroidal anti-inflammatory medications are typically required in adults depending on the specific underlying cause.

Future Diagnosis and Treatment

Research regarding diagnosis and treatment of EE continues to be a priority. A preliminary report by Aceves et al [62] described a survey instrument detailing abdominal symptoms in an attempt to distinguish EE from GERD in children. They found that early satiety and dysphagia were worse in patients with EE compared to patients with GERD. Research regarding biomarkers is being pursued for many diseases and EE is no exception. Konikoff et al [63] investigated biomarkers in a prospective, cross-sectional cohort of pediatric patients undergoing endoscopy for the possible diagnosis of EE. They obtained blood samples to measure eosinophil count, eosinophil-derived neurotoxin (EDN), eotaxin-1, -2, -3, and IL-5. Stool EDN levels were also determined. Levels of biomarkers were correlated with eosinophil density and disease activity. The authors found that blood eosinophil count, EDN, and eotaxin-3 levels were significantly correlated with eosinophil density on esophageal biopsy, as well as increased in active EE vs controls. A review of the current status of noninvasive markers of EE has recently been published [64].

IL-5 has been a therapeutic target for EE treatment. Stein et al. [65] reported the results of an open-label phase I/II safety and efficacy study of a humanized monoclonal anti-IL-5 IgG antibody (mepolizumab) in 4 adult patients with EE. The anti-IL-5 therapy was associated with a significant decrease in blood eosinophil levels, esophageal eosinophil counts, as well as improved clinical outcomes. Straumann et al. [66] recently reported the results of a randomized, placebo-controlled, double-blind trial of mepolizumab in 11 adult patients with severe EE. The authors found that active treatment decreased mean blood eosinophil levels as well as esophageal eosinophil concentrations (67% vs 25% in placebo group). A small subgroup experienced a decrease in symptoms. Mepolizumab may be effective for EGID. However, as with many new biological agents, the cost-effectiveness may only be established in the more severe patient groups.

Omalizumab is a humanized anti-IgE monoclonal IgG antibody. Foroughi et al [67] recently reported the results of an open-label investigation of anti-IgE (omalizumab) in 9 patients with EGID. Most patients had eosinophilic involvement of the esophagus, stomach, and duodenum. Omalizumab was associated with a significant decrease in blood eosinophil levels. There were no statistically significant changes in tissue eosinophil concentrations from baseline. Specific eosinophil levels decreased mildly in the duodenum and gastric antrum and body (not statistically significant, p = .074, .098, and 0.25, respectively), but increased slightly in the esophagus (again not significant, p = 0.47). Interestingly, there was a statistically significant decrease in symptom scores. Further research is necessary.

References

[1] Wershil, BK; Furuta, GT. Gastrointestinal mucosal immunity. *J. Allergy Clin. Immunol.* 2008;121:S380-383.

[2] Jang, MH; Kweon, MN; Iwatani, K; et al. Intestinal villous M cells: an antigen entry site in the mucosal epithelium. *Proc. Natl. Acad. Sci. U. S. A.* 2004;101:6110-6115.

[3] Miller, H; Zhang, J; KuoLee, R; et al. Intestinal M cells: the fallible sentinels? *World J. Gastroenterol.* 2007;14:1477–1486.

[4] Neutra, MR; Mantis, NJ; Kraehenbuhl, JP. Collaboration of epithelial cells with organized mucosal lymphoid tissues. *Nat. Immunol.* 2001;2:1004–1009.

[5] Niess, JH; Reinecker, HC. Dendritic cells: the commanders-in-chief of mucosal immune defenses. *Curr. Opin. Immunol.* 2006;22:354–360.

[6] Bilsborough, J; Viney, JL. Gastrointestinal dendritic cells play a role in immunity, tolerance, and disease. *Gastroenterology*. 2004;127:300–309.

[7] Marshall, JS. Mast-cell responses to pathogens. *Nat. Rev. Immunol.* 2004;4:787–499.

[8] Rothenberg, ME. Eosinophilic gastrointestinal disorders (EGID). *J. Allergy Clin. Immunol.* 2004;113:11-28.

[9] Chehade, M; Mayer, L. Oral tolerance and its relation to food hypersensitivities. *J. Allergy Clin. Immunol.* 2005;115:3-12.

[10] Michael, JG. The role of digestive enzymes in orally induced immune tolerance. *Immunol. Invest.* 1989;18:1049–1054.

[11] Kraehenbuhl, JP; Neutra, MR. Transepithelial transport and mucosal defense II: secretion of IgA, *Trends Cell. Biol.* 1992;2:170–174.

[12] Santos, LM; Al-Sabbagh, A; Londono, A; et al. Oral tolerance to myelin basic protein induces regulatory TGF-β-secreting T cells in Peyer's patches of SJL mice. *Cell Immunol.* 1994;157:439–447.

[13] Kim, PH; Kagnoff, MF. Transforming growth factor-β1 is a costimulator for IgA production. *J. Immunol.* 1990;144:3411–3416.

[14] Borsutzky, S; Cazac, BB; Roes, J; et al. TGF-β receptor signaling is critical for mucosal IgA responses. *J. Immunol.* 2004;173:3305–3309.

[15] Frossard, CP; Hauser, C; Eigenmann, PA. Antigen-specific secretory IgA antibodies in the gut are decreased in a mouse model of food allergy. *J. Allergy Clin. Immunol.* 2004;114:377–382.

[16] Sgouros, SN; Bergele, C; Mantides, A. Eosinophilic Esophagitis in adults: a systematic review. *Eur. J. Gastroenterol. Hepatol.* 2006;18:211-217.

[17] Furuta, GT; Liacouras, CA; Collins, MH; et al. Eosinophilic esophagitis in children and adults: A systematic review and consensus recommendations for diagnosis and treatment. *Gastroenterology.* 2007;133:1342-1363.

[18] Klein, NC; Hargrove, RL; Sleisenger, MH; et al. Eosinophilic gastroenteritis. *Medicine (Baltimore)* 1970;49:299-319.

[19] Talley, NJ; Shorter, RG; Phillips, SF; et al. Eosinophilic gastroenteritis: a clinicopathologic study of patients with disease of the mucosa, muscle layer, and subserosal tissues. *Gut.* 1990;31:54-58.

[20] Van Sickle, GJ; Powell, GK; McDonald, PJ; et al. Milk- and soy protein-induced enterocolitis: evidence for lymphocyte sensitization to specific food proteins. *Gastroenterology.* 1985;88:1915-1921.

[21] Noel, RJ; Putman, PE; Rothenberg, ME. Eosinophilic esophagitis. *N. Engl. J. Med.* 2004;351:940-941.

[22] Atwood, SEA; Smyrk, TC; Demeester, TR; et al. Esophageal eosinophilia with dysphagia. A distinct clinicopathologic syndrome. *Dig. Dis. Sci.* 1993;38:109-116.

[23] Fox, VL; Nurko, S; Furuta, GT. Eosinophilic esophagitis: It's not just kid's stuff. *Gastrointest. Endosc.* 2002;56:260-270.

[24] Straumann, A; Simon, HU. Eosinophilic esophagitis: escalating epidemiology? *J. Allergy Clin. Immunol.* 2005;115:418-419.

[25] Liacouras, CA; Wenner, WJ; Brown, K; et al. Primary eosinophilic esophagitis in children: successful treatment with oral corticosteroids. *J. Pediatr. Gastroenterol. Nutr.* 1998;26:380-385.

[26] Kerlin, P; Jones, D; Remedios, M; et al. Prevalence of eosinophilic esophagitis in adults with food bolus obstruction of the esophagus. *J. Clin. Gastroenterol.* 2007;41:356-361.

[27] Desai, TK; Stecevic, V; Chang, CH; et al. Association of eosinophilic inflammation with esophageal food impaction in adults. *Gastrointest. Endosc.* 2005;61:795-801.

[28] Ronkainen, J; Talley, NJ; Aro, P; et al. Prevalence of oesophageal eosinophils and eosinophilic oesophagitis in adults: the population-based Kalixanda study. *Gut.* 2007;56:615-20.

[29] Buckmeier, BK; Rothenberg, ME; Collins, MH. The incidence and prevalence of eosinophilic esophagitis. *J. Allergy Clin. Immunol.* 2008;121:S71.

[30] Cherian, S; Smith, NM; Forbes, DA. Rapidly increasing prevalence of eosinophilic oesophagitis in Western Australia. *Arch. Dis. Child.* 2006;91:1000-1004.

[31] Vanderheyden, AD; Petras, RE; DeYoung, BR; et al. Emerging eosinophilic (allergic) esophagitis: Increased incidence or increased recognition? *Arch. Pathol. Lab. Med.* 2007;131:777-779.

[32] Franciose, JP; Osman, J; Li, C; et al. Increasing incidence or increasing detection of esophageal eosinophils: A re-evaluation of esophageal biopsies from 1985 to 2005. *J. Allergy Clin. Immunol.* 2008;121:S71.

[33] Blanchard, C; Rothenberu, ME. Basic pathogenesis of eosinophilic esophagitis. *Gastrointest. Endoscopy Clin. North Am.* 2008;18:133-143.

[34] Mishra, A; Wang, M; Pemmaraju, VR; et al. Esophageal remodeling develops as a consequence of tissue specific IL-5-induced eosinophilia. *Gastroenterology.* 2008;134:204-214.

[35] Zuo, L; Mingler, M; Blanchard, FD; et al. IL-13 transgene-induced experimental eosinophilic esophagitis (EE) is associated with increased esophageal circumference and extensive angiogenesis. *J. Allergy Clin. Immunol.* 2008;121:S72.

[36] Blanchard, C; Wang, N; Stringer, KF; et al. Eotaxin-3 and a uniquely conserved gene-expression profile in eosinophilic esophagitis. *J. Clin. Invest.* 2006;116:536-547.

[37] Justinich, C; Pooni, A; Mak, N; et al. Esophageal epithelial cells as antigen-presenting cells? Implications for eosinophilic esophagitis. *J. Allergy Clin. Immunol.* 2008;121:S13.

[38] Chehade, M; Yershov, O; Shreffler, W; et al. Basophils are present in the esophagus of children with eosinophilic esophagitis. *J. Allergy Clin. Immunol.* 2008;121:S105.

[39] Blanchard, C; Wang, N; Rothenberg, ME. Eosinophilic esophagitis: Pathogenesis, genetics, and therapy. *J. Allergy Clin. Immunol.* 2006;118:1054-1059.

[40] Kelly, KJ; Lazenby, AJ; Rowe, PC; et al. Eosinophilic esophagitis attributed to gastroesophageal reflux: improvement with an amino acid-based formula. *Gastroenterology.* 1995;109:1503-1512.

[41] Markowitz, JE; Spergel, JM; Ruchelli, E; et al. Elemental diet is an effective treatment for eosinophilic esophagitis in children and adolescents. *Am. J. Gastroenterol.* 2003;98:777-782.

[42] Kagalwalla, AF; Sentongo, TA; Ritz, S; et al. Effect of six-food elimination diet on clinical and histologic outcomes in eosinophilic esophagitis. *Clin. Gastroenterol. Hepatol.* 2006;4:1097-1102.

[43] Spergel, JM; Andrews, T; Brown-Whitehorn, TF; et al. Treatment of eosinophilic esophagitis with specific food elimination diet directed by a combination of skin prick and patch tests. *Ann. Allergy Asthma Immunol.* 2005;95:336–343.

[44] Spergel, JM; Beausoleil, JL; Mascarenhas, M; et al. The use of skin prick tests and patch tests to identify causative foods in eosinophilic esophagitis. *J. Allergy Clin. Immunol.* 2002;109:363–368.

[45] Slingluff, MI; Haymore, BR; Foroughi, S; et al. Introducing percutaneous and patch testing to foods in a clinical setting for patients with eosinophilic esophagitis: A case series. *J. Allergy Clin. Immunol.* 2008;121:S104.

[46] Gottlieb, S; McEvoy, C; Winesett, D; et al. Food prick and atopy patch skin testing is necessary in children with eosinophilic esophagitis. *J. Allergy Clin. Immunol.* 2008;121:S104.

[47] Holbreich, M. A comparison of fresh vs commercial extracts for food testing. *J. Allergy Clin. Immunol.* 2008;121:S251.

[48] Ghaffari, G. Value of food patch testing in pediatric and adult patients with eosinophilic esophagitis and eosinophilic colitis. *J. Allergy Clin. Immunol.* 2008;121:S103.

[49] Simon, D; Straumann, A; Wenk, A; et al. Eosinophilic esophagitis in adults - no clinical relevance of wheat and rye sensitizations. *Allergy*. 2006;61:1480-1483.

[50] Gonsalves, N; Ritz, S; Yang, G; et al. A prospective clinical trial of allergy testing and food elimination diet in adults with eosinophilic esophagitis (EE). *Gastroenterology.* 2007;132:A6.

[51] Mishra, A; Hogan, SP; Brandt, EB; et al. An etiological role for aeroallergens and eosinophils in experimental esophagitis. *J. Clin. Invest.* 2001;107:83-90.

[52] Fogg, MI; Ruchelli, E; Spergel, JM. Pollen and eosinophilic esophagitis. *J. Allergy Clin. Immunol.* 2003;112:796-797.

[53] Vannelli, P; McGeady, SJ; Garola, R; et al. Comparison of the seasonal and geographic distribution of children with eosinophilic esophagitis. *J. Allergy Clin. Immunol.* 2008;121:S104.

[54] Zuo, L; Rothenberg, ME. Gastrointestinal eosinophilia. Immunol. *Allergy Clin. North Am.* 2007;27:443-455.

[55] Ahmad, M; Soetikno, RM; Ahmed, A. The differential diagnosis of eosinophilic esophagitis. *J. Clin. Gastroenterology*. 2000;30:242-244.

[56] Liacouras, CA; Spergel, JM; Ruchelli, E; et al. Eosinophilic esophagitis: a 10-year experience in 381 children. *Clin. Gastroenterol. Hepatol.* 2005;3:1198-1206.

[57] Ferguson, DD; Foxx-Orenstein, AE. Eosinophilic esophagitis: an update. *Dis. Esophagus.* 2007;20:2-8.

[58] Assa'ad, AH; Putman, PE; Collins, MH; et al. Pediatric patients with eosinophilic esophagitis: An 8-year follow-up. *J. Allergy Clin. Immunol.* 2007;119:731-738.

[59] Potter, JW; Saeian, K; Staff, D; et al. Eosinophilic esophagitis in adults: and emerging problem with unique esophageal features. *Gastrointest. Endosc.* 2004;59:355-361.

[60] Straumann, A; Spichtin, H-P; Grize, L; et al. Natural history of primary eosinophilic esophagitis: A follow-up of 30 adult patients for up to 11.5 years. *Gastroenterology.* 2003;125:1660-1669.

[61] Croese, J; Fairley, SK; Masson, JW; et al. Clinical and endoscopic features of eosinophilic esophagitis in adults. *Gastrointest. Endosc.* 2003;58:516-522.

[62] Aceves, SS; Arii, B; Dohil, M; et al. Prospective analysis of an abdominal symptom scoring tool's efficacy in the clinical distinction of pediatric eosinophilic esophogitis from gastroesophageal reflux disease. *J. Allergy Clin. Immunol.* 2008;121:S70.

[63] Konikoff, MR; Blanchard, C; Kirby, C; et al. Potential of blood eosinophils, eosinophil-derived neurotoxin, and eotaxin-3 as biomarkers of eosinophilic esophogitis. *Clin. Gastroenterol. Hepatol.* 2006;4:1328-1336.

[64] Gupta, S. Noninvasive markers of eosinophilic esophogitis. *Gastrointest. Endoscopy Clin. North Am.* 2008;18:157-167.
[65] Stein, M; Collins, MH; Villanueva, JM; et al. Anti-IL-5 (mepolizumab) therapy for eosinophilic esophagitis. *J. Allergy Clin. Immunol.* 2006;118:1312-1319.
[66] Straumann, A; Conus, S; Kita, H; et al. Mepolizumab, a humanized monoclonal antibody to IL-5, for severe eosinophilic esophogitis in adults: A randomized, placebo-controlled double-blind trial. *J. Allergy Clin. Immunol.* 2008;121:S44.
[67] Foroughi, S; Foster, B; Kim, NY; et al. Anti-IgE treatment of eosinophil-associated gastrointestinal disorders. *J. Allergy Clin. Immunol.* 2007;120:594-601.

In: Food Allergies: New Research
Editor: Carrie M. Chesterton, pp. 189-201
ISBN 978-1-60456-978-0

Chapter 7

Preventing and Responding to Food Anaphylaxis in School Settings

Genevieve H. Hay[1,*], Thomas B. Harper III[2,†] and Peachey M. Trudell[3,‡]

[1]Elementary Education, School of Education, Health, and Human Performance, College of Charleston, Charleston, SC 29424, USA
[2]Department of Pediatrics, Medical University of South Carolina, Charleston, SC 29425, USA
[3]College of Charleston, Early Childhood Education, School of Education, Health, and Human Performance, College of Charleston, Charleston, SC 29424, USA

Abstract

The chapter reviews current research regarding the prevalence of severe food allergies in school age children and the most effective preventative and treatment practices to be implemented in schools. Common concerns revealed in the literature center around the risks involved when school personnel lack knowledge and awareness of anaphylaxis and the necessary emergency protocols to put in place in the event of a reaction. The authors examine the critical role of epinephrine in the early treatment of anaphylaxis and the urgent need for school nurses, physicians, parents, and educators to put effective protocols in place. To better ensure that epinephrine be administered promptly and that care is carefully coordinated with emergency personnel, school nurses need to be the coordinator of care. Analysis of recent studies regarding the school nurse's role in the development of emergency medical plans for children with special health care needs (CSHCN) reveals the need to have one full-time school nurse per school to better assure access to prompt, quality care. It is vital for school nurses to actively participate in the development of Individualized Education Plans (IEPs) or Individual Health Plans (IHPs) for students identified to have specific health related disabilities and be designated as the primary individual in charge of implementation of these plans. Across the United

* hayg@cofc.edu
† tbharperiii@comcast.net
‡ pmlyne@edisto.cofc.edu

States there are dramatic differences in state and local policies impacting school nurse and student ratios which, in turn, can have a significant impact upon management and care provided to students with special medical needs. Recommendations for standards of care for students with severe food allergies will be discussed.

An adverse reaction to a food can be due to a variety of mechanisms. *Food intolerance* is a term used to describe an abnormal or exaggerated physiologic response to an ingested food or food additive (there are more than 3,000 used in processed foods today). These can include lactose intolerance, reactions to food toxins or bacteria, or reactions to food additives such as MSG or sulfites. These reactions are usually not serious but may recur and are a concern to families. The term *food allergy* is reserved for an abnormal or exaggerated immunologic response to foods. These types of reactions typically are acute and can lead to anaphylaxis.

Anaphylaxis is a potentially life-threatening allergic or allergic-like reaction resulting from exposure to a substance to which an individual has become sensitized. Anaphylaxis usually occurs suddenly after exposure with an allergy-causing substance although symptoms may be delayed 1-3 hours after exposure. The bodily systems that can be involved include skin (itch, hives, swelling, flushing, sweating), gastrointestinal (vomiting, diarrhea, cramping), respiratory (runny nose, sneezing, wheezing, cough, hoarseness, shortness of breath), cardiovascular (tachycardia, hypotension, shock, syncope, cyanosis) and neurological (anxiety, headache, seizure, feeling of "impending doom"). Often, the individual's acute presentation may not contain all of the above symptoms; although, at least two systems must be involved to be considered anaphylaxis. Differing temporal patterns of anaphylaxis exist and can include acute (symptoms resolve within hours of treatment), biphasic (recurrence of symptoms 1-4 hours after initial event), and protracted (unrelenting anaphylactic symptoms for up to 24 hours) (Simons, 2008).

A variety of triggers can lead to the various clinical manifestations of anaphylaxis. In the school setting, the most common trigger for anaphylaxis is food, with major culprits being nuts (peanuts and tree nuts), cow's milk, eggs, soy, and wheat. Anaphylaxis to fish and crustaceans (shrimp, crab, and lobster) can occur in children, but is more common in adults. Other occasional causes of anaphylaxis in school age children include reaction to members of the Hymenoptera order of insects (fire ant, bee, wasp, and yellow jacket), exposure to latex, reactions to medication (particularly antibiotics and aspirin-related medications), and exercise (occasionally related to prior specific food ingestion). Rarely, anaphylaxis can be due to unknown factors in spite of extensive medical evaluation. Anaphylaxis can also occur in early childhood, usually due to foods such as egg or cow's milk, antibiotics, or latex in baby bottle nipples, pacifiers, or toys.

The pathophysiologic mechanisms of anaphylaxis involve the release of mediators or chemicals from cells like mast cells or basophils throughout the body after exposure to an allergen. It is the body's response to these mediators that causes the various symptoms of anaphylaxis. Food exposure triggering anaphylaxis usually is the result of ingestion of the food although there are reports of skin contact or inhalation causing systemic reactions (Sicherer, Furling, DeSimone, & Sampson, 1999).

The incidence of anaphylaxis has been difficult to quantify due to differing criteria for the definition of anaphylaxis. A recent commentary by the American College of Allergy, Asthma, and Immunology (ACAAI) estimate a lifetime prevalence of approximately 0.05%

to 2.0% (Lieberman et al., 2006). Using the 1999 United States population of 272 million, Neugut and colleagues estimated that between 3.3 and 43 million Americans were at risk for an anaphylactic reaction because of food, medication, latex, or insect sting, including 1500 fatal reactions (Neugut, Ghatak, & Miller, 2001). Food anaphylactic deaths are probably underestimated as autopsy findings in fatal food-related anaphylaxis may be negative or misdiagnosed as acute asthma or acute coronary syndrome (Klein & Yocum, 1995). Reviews of these fatal reactions have consistently shown that the greatest risk is in adolescents and young adults (58% of deaths between ages of 13 and 30), and that an injectable source of epinephrine was usually not available (Bock, Muñoz-Furlong, & Sampson, 2001; Bock, Muñoz-Furlong, & Sampson, 2007). Adolescents are at particular risk as they are more removed from parental supervision and are trying very hard to "fit in," sometimes by denying or ignoring their specific food sensitivities with occasional tragic consequences (Sampson, Muñoz-Furlong, & Sicherer, 2006). Through the efforts of the Food Allergy and Anaphylaxis Network (FAAN), national attention has recently been focused on ensuring the safety of food allergic children in the school system.

The medical treatment of anaphylaxis ultimately involves an aggressive effort by a team of physicians in an emergency facility to stabilize respiratory and cardiovascular systems; however, the single most important and life saving treatment involves the early use of injected epinephrine and rapid transport of the involved student to a medical facility (Simons, 2004). Currently, avoidance of the offending food and aggressive treatment of "food accidents" are the only available therapies for the growing number of children with food allergies. Research is currently underway to develop more active forms of therapy such as food desensitization and various forms of immunomodulation to reduce patient sensitivity to food allergens (Nowak-Wegryzyn, 2003; Valenta & Kraft, 2002).

FAMILIES: ANXIETY AND COPING

Over the years, there has been a void in the research related to the impact of living with food allergies. Recent studies have examined the emotional impact of being diagnosed and learning to live with life-threatening food allergies upon children and their families. These studies expose the need for close collaboration of a multidisciplinary group of professionals, including physicians, social workers, psychologists, school nurses, school administrators, and teachers, who should assist families with coping, devise emergency plans, and address, with the families, the physical and emotional needs of their children with severe food allergies.

A 2005 psychosocial study of families (Mandell, Curtis, Gold, & Hardy, 2005) examined aspects of coping and adjusting to life with severe food allergies. In the study, the authors utilized qualitative methodology in which they interviewed 17 families where developmental cycles of coping and adapting were revealed. The researchers found remarkably consistent coping patterns in families living with the potential for anaphylaxis in a child. Upon diagnosis, families tended to be hyper-vigilant where they sought information about the condition and avoidance measures. Over time, families' levels of anxiety seemed to peak upon accidental exposure to an allergen, as their children moved through specific developmental milestones, and as they discovered new research. This study, as well as others, indicate that there is an optimal level of anxiety, referred to as "The Goldilocks

Principle," where families develop a level of anxiety which is "just right" as they learn to live day to day with anaphylaxis (Mandell et al., 2005; Mandell, Curtis, Gold, & Hardy, 2002; Primeau et al., 2000; Avery, King, Knight, & Hourihane, 2003) where the family is neither hyper-vigilant nor passive. It should be noted that while similarities exist in the families' ability to cope to chronic illnesses, like asthma, families learning to live with anaphylaxis were unique in that their adjustment did not tend to be linear, instead it was cyclical. It is important to understand that families tend to go through "out-of-control" phases where levels of anxiety are heightened after periods of confidence with their ability to cope. Information found on coping patterns should provide implications for professionals as they work with families. Medical professionals should make efforts to provide pertinent information and assist families in developing support systems (Mandell et al., 2005).

Initially, some families have difficulty grasping the seriousness of the diagnosis; whereas other families have been described as being extremely anxious. Many of the families described a heightened level of anxiety at this time due to being provided with insufficient information by physicians. However, many families conveyed that upon visits to allergists they received more support. Families found that, in general, allergists tended to be more informative and would provide education and suggestions for support. At this critical time, medical professionals should be aware of the need to provide detailed information about the nature of the allergy, precautions to take, instructions for administering medications, and where to find support (Mandell et al., 2005).

As children moved through specific developmental stages, parents and children were found to have periods of intensified anxiety, especially at times when children became more independent. Parental and child anxiety was highest for children ages six to11. At this time, children become more autonomous and cognizant of the severity of their allergies, yet they are not yet able to fully protect themselves. Logically, as children reach school age, many families were found to become increasingly anxious about their child's safety at school where their children would encounter numerous staff members and other children, and be exposed to food in various school settings. Interestingly, many parents did not express undue concern about the level of caution that their adolescent children possess. Instead, parents were most concerned about whether their adolescent carried their epinephrine at all times (Mandell et al., 2005). Findings from the 2005 Mandell study, as well as others, reveal that adolescents may be particularly at high risk since they do not consistently carry their epinephrine, they tend to engage in risk-taking behaviors, and because they tend to be less inclined to monitor ingredients (Mandell et al., 2005; Sampson, Muñoz-Furlong, & Sicherer, 2006; Bock et al., 2001; Bock et al., 2007).

In 2006, risk taking behaviors among 174 adolescents ranging in age from 13 to 21 who have documented food allergies were investigated (Sampson, Muñoz-Furlong, & Sicherer, 2006). The researchers found that only 61% of the adolescent participants stated that they "always" carried their epinephrine at all times. Depending upon specific circumstances, dramatic differences existed in whether or not the surveyed group of adolescents carried their epinephrine. The majority of the adolescents carried their epinephrine when traveling or when at restaurants. However, rates of carrying epinephrine decreased significantly when participating in social activities, sporting activities, and when wearing tight clothing (Sampson et al., 2006). In a previous study conducted in 2005, McIntyre and colleagues found that even though many adolescents are allowed to carry epinephrine, some adolescents were not carrying their medications (McIntyre, Sheetz, Carroll, & Young, 2005). It was, therefore, recommended that supplies of epinephrine be maintained by schools. Additionally, it should be noted that the Board of Directors of the American Academy of Allergy, Asthma, and Immunology (AAAAI, 1998) recommended that all students have assistance administering epinephrine due to potential difficulties that an individual may experience when self-administering the drug while having a severe allergic reaction.

A major concern found in the 2005 Mandell study was that of enlisting support from individuals involved in a child's life. Families reported difficulty securing cooperation of others due to failure to comprehend the seriousness of the problem or due to fearfulness. This reinforces the apprehension that many families experience when they have to entrust their children with other caretakers. It also helps to place emphasis on the need for understanding the condition and attempt to understand any coping difficulties that the family may be experiencing.

Overall, the Mandell study revealed an increase in vigilance among most families after accidental exposure to a food allergen. Recommendations from the study include the establishment of a multidisciplinary approach of physicians, social workers, and school personnel, especially as families move through typical crisis stages and as appropriate emergency plans are devised and carefully implemented. Multidisciplinary teams should seek information from families regarding their coping strategies and provide assistance to families in their attempts to develop an adaptive level of coping (Mandell et al., 2005).

PREVENTION AND POLICIES

Given the increasing incidence of food allergies in children, the need exists for preparation of school personnel by the school nurse (Weiss, Muñoz-Furlong, Furlong & Arbit, 2004). Children are particularly at-risk of having anaphylactic reactions since 6-8% of them have food allergies (Sicherer, Furlong, DeSimone, & Sampson, 2001; Weiss, 2004). Peanut and tree nut allergies are of particular concern since these allergies are usually not outgrown and since a number of severe and fatal reactions occurred in schools. In a study of six fatal and four near-fatal reactions, five of the fatalities were from peanut or nut allergies with four of the fatalities caused by reactions occurring in schools (Sampson, 1992).

Alarmingly, similar studies conducted within the past decade in the United Kingdom and the United Studies indicate that peanut allergy has doubled in children (Grundy, Matthews, Bateman, Dean, & Arshad, 2002; Sicherer, Muñoz-Furlong & Sampson, 2003). Grundy and colleagues found a 2-fold increase in peanut allergy and a 3-fold increase in peanut sensitization. In 2002, through a randomized national telephone survey, Sicherer and colleagues (2002) documented that the number of peanut allergy cases in children had steadily risen from 0.4% in 1997 to 0.8%. Possible reasons for the increase in children may be related to the process of roasting peanuts to make peanut butter, early exposure to peanut when the immune system is immature, and the use of topical ointments containing peanut (Grundy et al., 2002; Sicherer et al., 2003; Sicherer & Malloy, 2005).

In a 2001 parent telephone interview study of children with a history of allergic reactions in school, Sicherer and his colleagues found that 16% of the children had reactions in school or child care. In many reactions, the allergic reaction was due to craft projects involving peanut products (e.g., bird feeders) and celebrations that included baked goods or shared food. Craft projects and cooking activities can lead to high levels of exposure for some allergic individuals. There was a nurse on site for only 23% of these reactions. The first adult to recognize symptoms was the teacher (59%), parents when picking their child up from school (32%), and other school personnel in the remaining cases. In 60% of the cases, parents were notified. Emergency plans were in place for only 33% of the students and followed only 73% of the time. Training of epinephrine administration occurred in 25 of the 29 cases prior to its use (Sicherer, Furlong, DeSimone, & Sampson, 2001).

A recent update on peanut allergy by Sicherer extends recommendations for allergic food avoidance in school made by the American Academy of Allergy in 1998 (Sicherer, 2002). These suggestions include increased supervision during meals and snacks, no food or utensil sharing, cleaning of tables and toys, substitution of causal foods during craft, cooking and science projects, hand washing before/after food handling, provision of safe substitution foods, ingredient lists for any foods brought from home, and instruction of staff on issues including careful label reading, cross-contamination, and technical/scientific word(s) for the food(s) (Sicherer, 2002).

Despite the AAAAI's 1998 Position Statement, schools have fallen short in their responsibility to care for students who have the potential of having anaphylactic reactions. A 2005 study in 109 Massachusetts schools sought to determine the incidence of anaphylaxis in schools, the availability and use of epinephrine, the training of school personnel, and the existence of emergency plans. Results from the study revealed that epinephrine was infrequently used or was unavailable, emergency plans were not always in place, and staff training had not consistently occurred. In cases where epinephrine was administered, the average time from onset of symptoms to administration was 10 minutes with a range up to 75 minutes (McIntyre et al., 2005).

McIntyre and colleagues made several recommendations based upon the findings from their study (2005). First, epinephrine should be on-site and used immediately when anaphylactic symptoms are present. In some cases, there was no supply of epinephrine or physician's orders for its use. Second, comprehensive staff training is needed in schools. Many schools did not train the staff in prevention strategies and reaction response. Third, all individuals treated with epinephrine should be transported to the hospital immediately due to the possibility of biphasic reactions. In this study, nine (8%) individuals were not transported to an emergency facility. Fourth, as supported in the McIntyre study and by the National Association of School Nurses (NASN), children with special health care needs (CSHCN) or children with chronic medical conditions which require medical services that are not typical of other children, need to have emergency care plans (ECPs) in place. The McIntyre study found that ECPs existed in 92% of the cases; however, the schools emergency response system was implemented in only 62% of the cases.

A significant finding in the McIntyre study was that allergic reactions occurred in 24% of individuals who had not previously experienced allergic reactions. Similarly, the 2001 Sicherer study found that first time reactions to peanuts occurred in 25% of students while at school. Additional recommendations from the McIntyre study include the implementation of physician signed protocols to authorize the administration of epinephrine by the school nurse to anyone experiencing anaphylactic symptoms. Furthermore, back up supplies of epinephrine should be available in schools and stored in multiple sites. If the nurse is unavailable, protocols should establish that school personnel be trained to recognize symptoms of anaphylaxis and call emergency personnel for treatment (McIntyre et al., 2005).

THE ROLE OF SCHOOL NURSES

In their position statement, the National Association of School Nurses (2006b) outlines school nursing policies and procedures. In 1975, Congress enacted the Education for All Handicapped Children Act (EHA) to ensure free and appropriate public education for students with disabilities, including students with health related conditions. The act was amended and renamed in 1990 as the Individuals with Disabilities Education Act (IDEA) and has since been amended in 2004 and 2007. The 2004 reauthorization includes nursing services as a related service to help ensure that students' health needs are met and that these students are educated in the least restrictive environment. Students with health related disabilities are to be provided medical services or supervised medical services by a Registered Professional School Nurse. Two US Supreme Court decisions, Irving Independent School District v Tatro (1984) and Cedar Rapids Community School District v Garret F. (1999) gave impetus to IDEA's requirements for students with health related disabilities to be provided nursing services, if deemed necessary, as a part of their Individualized Education Plan (IEP) in order for the students to access and benefit from their educational programs (Thies, 1999). If students are not served under IDEA, they may be provided services through the Section 504 of the Vocational Rehabilitation Act (1973) where reasonable accommodations must be provided to students with a disability or a substantial limitation to a major life activity, including health and learning needs. 504 plans provide reasonable accommodations which may include school nurses services, such as medication administration as well as other health-related procedures.

Through IDEA and 504, students with health related disabilities have detailed educational plans stipulated through an Individualized Education Plan (IEP) and detailed medical plans in an Individualized Health Care Plan (IHP) or in a 504 accommodation plan. An IHP is a detailed school health plan addressing all of a student's health care needs; whereas an ECP (Emergency Care Plan) is the specific emergency response protocol that school nurses should develop along with the multidisciplinary team and the child's family for the emergency treatment of a child experiencing a medical emergency, such as an anaphylactic reaction in and outside of the school facility. To better ensure effective ECPs, school nurses need to be the coordinator of care where they assist in the planning of emergency responses which are carefully coordinated with emergency personnel (National Association of School Nurses [NASN], 2006b; Sheetz et al., 2004).

In 2005, Olympia and colleagues randomly selected 1000 school nurses to survey in order to determine school preparedness to respond to medical emergencies. Sixty-nine percent of the forms were completed and returned. In association with other studies related to emergency preparedness for students with chronic illnesses, the researchers obtained similar findings. For instance, 86% of the schools had medical emergency response plans (MERPs) or ECPs in place. However, only 35% of the schools practiced the plan. Thirteen percent of the schools did not designate personnel to make medical emergency decisions. In 205 schools without a full-time school nurse, 17% did not have MERPs, 17% did not designate personnel to make medical emergency decisions and 72% did not have effective communication systems. As further support for emergency care planning, the researchers concluded that schools would be better prepared to deal with life-threatening emergencies by practicing their emergency care plans several times per year, by having effective campus communication systems, and by ensuring that personnel are assigned and practice designated roles in response to a medical emergency (Olympia, Wan, & Avner, 2005).

In 2004, Weiss and his colleagues conducted a nationwide study of 400 school nurses in an attempt to determine the impact of food allergies on school nursing practices. Forty-four percent of the nurses reported an increase in the number of students with food allergies. Overall, the nurses reported a mean of 9.9 students in their schools with food allergies. When asked to measure the challenge of nursing children with special health care needs (CHCSN) on a scale of 1 ("not a challenge at all) to 10 ("a significant challenge"), 29% of the nurses rated food allergies at either 8, 9 or 10, or similar to how they rated conditions such as diabetes. In comparison to other health related problems, 87% of the nurses indicated that food allergies in children is a somewhat or a very serious concern. Seventy-one percent of the nurses reported that caring for students with food allergies is very burdensome (Weiss et al., 2004).

The Weiss study (2004) revealed concerns regarding a significant number of severe food allergic reactions occurring in schools and inadequate school responses to allergic reactions. Inconsistencies in staff training and standardized training plans were prevalent. The researchers found that 87% of schools had emergency care plans (ECPs) in place for students with food allergies, staff was trained in 78% of the time, 73% of the schools had food trading restrictions, and 81% of the schools developed emergency plans for field trips. In addition to inconsistent training programs, it was found that parents, rather than school nurses, conducted staff training in 47% of the cases.

Weiss and colleagues (2004) found that many of the school nurses were unaware of published resources, available at no cost, such as the Massachusetts Department of Education 2002 plan, *Managing Life Threatening Food Allergies in Schools* (Sheetz et al., 2004), and *The School Food Allergy Program* from the Food Allergy and Anaphylaxis Network (FAAN) as well as additional support and training materials that FAAN provides. Seventy-four percent were unaware of FAAN altogether. As a result, nurses were continually "reinventing the wheel" when developing food allergy staff training programs.

Furthermore, the Weiss study (2004) revealed that 13% of nurses were part time and 32% of the nurses served more than one school. This raises concern about who will coordinate care and locate epinephrine in the event of a reaction when the nurse is not present. As previously stated, delays in the administration of epinephrine have been cited as a factor in the deaths of students. Like Weiss and colleagues, other researchers have recommended that effective emergency response teams be put in place where multiple staff members are trained to efficiently recognize and treat a reaction and that epinephrine storage be readily known. Even when a nurse is on-site, there is the possibility that the nurse may be treating another student or that the nurse may be in a remote part of the campus (McIntyre et al., 2005; Sicherer et al., 2001). The findings from the Weiss study (2004) and others imply the need for national standards of care, including among other recommendations, a full-time registered school nurse at each school (Weiss et al., 2004; Sheetz et al., 2004; McIntyre et al., 2005; Sicherer et al., 2001; Brener, Wheeler, Wolfe, Vernon-Smiley, & Caldart-Olson, 2007; NASN, 2006a).

To help schools more safely manage students with severe food allergies, a bill was proposed in the United States Senate by Senator Chris Dodd of Connecticut. The bill, entitled "The Food Allergy and Anaphylaxis Act of 2007," was proposed on April 26, 2007 and is currently under review. If passed, the purpose of the act is to provide "consistent, standardized" guidelines to assist schools in the management of food allergy and anaphylaxis. In addition, schools will be able to obtain grants to assist in the development of comprehensive, standardized guidelines (S. 1232, 2007).

In an effort to effectively implement a multidisciplinary approach in schools, the school nurse should serve as the coordinator of care and planning for students with significant medical problems. In that capacity, the nurse, along with the parents, the school staff, and the child's physician, constructs the IHP, including an Emergency Care Plan (ECP) or emergency treatment plan. As health care coordinators, school nurses have access to all members of the school, the parents, and physicians (NASN, 2004). Even though faculty and staff should be trained to administer life-saving medications and cardiopulmonary resuscitation, school nurses are uniquely qualified to perform more sophisticated health services (Brener et al., 2007).

Analysis of recent studies regarding the school nurse's role when caring for children with special health care needs (CSHCN) reveals the need to have one full-time school nurse per school to better assure access to prompt, quality care. Across the United States there are dramatic differences in state and local policies impacting school nurse-to-student ratios which, in turn, can have a significant impact upon management and care provided to students with special medical needs (Guttu, Engelke, & Swanson, 2004). Historically, the National Association of School Nurses (NASN) and the federal government have recommended a school nurse to student ratio of 1:750. In schools where there are more significant numbers of students with disabilities and chronic medical conditions, it is recommended that there be better nurse-to-student ratios. More recently, NASN has recommended that one professionally trained (preferably nurses with a baccalaureate degree), full-time school nurse be in every school building to meet the medical needs of students (NASN, 2006a; Brener et al., 2007).

In 2006, The Centers for Disease Control and Prevention conducted its sexennial School Health Policies and Programs Study (SHPPS) to ascertain information about specific health services provided to students in all 50 states. Among their findings were that 86.3% of schools had a part-time nurse, 35.7% had a full-time nurse, and 45.1% of all schools had nurse-to-student ratios of 1:750. SHPPS recommended that the breadth of school nursing services be increased. A critical initial step would be to improve the nurse-to-student ratio in schools. In order for more encompassing reforms to take place, school districts will need policy support through legislation and funding (Guttu et al., 2004; Brener et al., 2007).

School Plans and Emergency Response

Proposals for managing food allergy in the school setting have been published based upon the recommendations of the Board of Directors of the American Academy of Allergy, Asthma, and Immunology (AAAAI, 1998). A summary of treatment protocol suggestions recommended by Sicherer for food allergic children in schools include physician-prescribed treatment protocols in place with periodic review, epinephrine available for potentially life-threatening reactions (readily available, not locked), and education of supervising personnel on (a) recognizing the signs of an allergic reaction, (b) techniques of medication administration, and (c) basic first aid and resuscitation (AAAAI, 1998; Sicherer, 2002). The development and implementation of an action protocol by the school multidisciplinary team for a specific child's food allergy is imperative and should involve input from the student, parents and their physician, teachers, and the school nurse (NASN, 2006b).

The American Academy of Pediatrics Committee on School Health recommends that schools be prepared to treat anaphylaxis in students (AAAAI, 1998; McIntyre et al., 2005; Sampson, 2004). In addition to IDEA and 504 requirements (NASN, 2004), public schools must comply with the Americans with Disabilities Act (ADA) which mandates accommodations for students with allergies (AAAAI, 1998). ADA guidelines indicate that accommodations be made to enable students with allergies to participate in all school activities. In addition, effective treatment protocols should be developed by physicians and efficiently implemented for students while ensuring that medications are always accessible. (AAAAI, 1998; NASN, 2004; Muñoz-Furlong, 2004)

As previously indicated, parents of a child with a life-threatening food allergy should provide written documentation of the student's allergy and appropriate treatment protocols from their physician. The documentation should assist school nurses in coordinating the school's emergency response team in the development of effective and efficient emergency treatment plans (Weiss et al., 2004). At least two automatic epinephrine injection devices should be supplied by the parents for all children with previous food reactions (AAAAI, 1998; McIntyre et al., 2005; Hay, Harper, & Moore, 2006). These devices should accompany the child on class trips or any outings (AAAAI, 1998; McIntyre et al., 2005). An individual child's emergency plan should delegate who carries and administers the medication. With physician approval, some mature and responsible students may be allowed to carry prescribed epinephrine while at school. (Weiss et al., 2004) Nevertheless, it is imperative that staff be trained and backup of medications be available (Weiss et al., 2004; Sicherer et al., 2001; Sampson et al., 2006). Plans should be made for calling emergency medical services, parents, and physician, who will stay with the child having the reaction.

School planning should include procedures for alerting the child's teachers and a prominent notation should be made on the child's health record. Staff training should also incorporate being trained to administer CPR and first aid. Phone numbers of the nearest medical facility and the estimated response time of a local EMS team should be determined (Hay et al., 2006).

When responding to anaphylactic reactions, school personnel should begin with recognition of the symptoms and prompt administration of epinephrine. Treatment of anaphylactic reactions should be sequential. In this order, calls should be made to emergency medical services (911), the child's parents, and the child's physician (if phone number is available). Additional medication administration should be deferred to EMS personnel or to physicians in an emergency room. Children experiencing anaphylactic reactions should not be transported to an emergency facility by school personnel under any circumstances unless emergency transport services are unavailable in the community. Immediate administration of epinephrine at the school and further treatment by emergency medical services at the site and during transport are essential aspects of therapy to reverse a potentially life-threatening anaphylactic reaction (Hay et al., 2006).

Awareness and Inclusion of Students

The social and psychological impact of food allergies on students, families, teachers, and school personnel should be considered. The presence of food in the cafeteria, classroom or school events can be stressful for food allergic individuals. Avoidance is the key to managing food allergies. Thorough precautions should be put in place to prevent allergic reactions and to prevent stigma and isolation of children with severe allergies (AAAAI, 1998). To promote inclusion and acceptance, teachers can conduct allergy awareness lessons to foster appropriate social interactions among students. Training kits and awareness materials are available to parents, children, and schools from The Food Allergy Network (www.foodallergy.org), a national food allergy support group. Teachers should ensure that their lessons are age appropriate and include accurate information regarding food allergies, effects of allergies, and emergency plans. Through teacher and school nurse directed lessons, acceptance and understanding of those with food allergies can occur. Understanding, empathy, and friendship can be promoted and misunderstanding and potentially dangerous ridicule can be prevented. Pertinent and accurate knowledge of food allergies can better assure the prevention of allergic reactions and can help maintain a safe school environment for students with severe food allergies. (Salend, 2008; Hay et al., 2006)

Conclusion

The diagnosis of life-threatening food allergies in children has tremendous social, emotional, and medical implications for families. Pediatric patients and their families need guidance in their quest to obtain accurate and relevant support from a multidisciplinary group of professionals (NASN, 2006b). Families must learn to effectively manage their child's food allergy and develop an optimal level of coping (Mandell et al., 2005). In addition, appropriate medications must be prescribed and families must be taught how to successfully handle medical emergencies (Bock et al., 2001).

When considering the possibility of potential life-threatening food allergic reactions in school children, it is imperative that teachers, administrators, school nurses, social workers and outside medical personnel collaborate to ensure proper prevention and treatment of children with severe food allergies. School personnel must obtain the necessary, documented medical information from parents. If school personnel are open, knowledgeable, and understanding, they can successfully develop and implement effective health plans to prevent food reactions and develop emergency action plans to implement when emergency life-threatening food reactions occur.

References

AAAAI Board of Directors (1998). Anaphylaxis in schools and other child-care settings. *Journal of Allergy and Clinical Immunology*, 102(2), 173-176.

Avery, N.J., King, R.M., Knight, S., & Hourihane, JO'B. (2003). Assessment of quality of life in children with peanut allergy. *Pediatric Allergy and Immunology*, 14, 378-382.

Bock, S.A., Muñoz-Furlong, A., & Sampson, H.A. (2001). Fatalities due to anaphylactic reactions to foods. *Journal of Allergy and Clinical Immunology*, 107, 191-193.

Bock, S.A., Muñoz-Furlong, A., & Sampson, H.A. (2007). Further fatalities caused by anaphylactic reactions to food, 2001-2006. *Journal of Allergy and Clinical Immunology*, 119(4), 1016-1019.

Brener, N.D., Wheeler, L., Wolfe, L.C., Vernon-Smiley, M., & Caldart-Olson, L. (2007). Health services: Results from the school health policies and programs study 2006. *Journal of School Health*, 77(8), 464-485.

Food Allergy and Anaphylaxis Management Act of 2007, S. 1232, 110th Cong. (2007).

Grundy, J.S., Matthews, S., Bateman, B., Dean, T., & Arshad, S.H. (2002). Rising prevalence of allergy to peanut in children: Data from 2 sequential cohorts. *Journal of Allergy and Clinical Immunology*, 110(5), 784-789.

Guttu, M., Engelke, M.K., & Swanson, M. (2004). Does the school nurse-to-student ratio make a difference? *Journal of School Health*, 74(1), 6-9.

Hay, G.H., Harper, T.B., & Moore, T.G. (2006). Assuring the safety of severely food allergic children in school. *Journal of School Health*, 76(9), 479-481.

Klein J.S. & Yocum, M.W. (1995). Underreporting of anaphylaxis in a community emergency room. *Journal of Allergy and Clinical Immunology*, 95(2). 637-638.

Lieberman, P., Camargo, C.A., Bohlke, K., Jick, H., Miller, R.L., Sheikh, A., & Simons, F. (2006). Epidemiology of anaphylaxis: Findings of the ACAAI epidemiology of anaphylaxis working group. *Annals of Allergy, Asthma and Immunology*, 97(5). 596-602.

Mandell, D., Curtis, R., Gold, M., & Hardie, S. (2002). Families coping with a diagnosis of anaphylaxis in a child: A qualitative study of informational and support needs. *ACI International*, 14(3), 96-101.

Mandell, D., Curtis, R., Gold, M., & Hardie, S. (2005). Anaphylaxis: How do you live with it? *Health & Social Work*, 30(4), 325-335.

McIntyre, C.L., Sheetz, A.H., Carroll, C.R., & Young, M.C. (2005). Administration of epinephrine for life-threatening allergic reactions in school settings. *Pediatrics*, 116, 1134-1140.

Muñoz-Furlong, A. (2004). Patient's perspective and public policy regarding anaphylaxis. In G. Bock, & J. Goode (Eds.), *Anaphylaxis* (pp. 265-275). Chichester, UK: Wiley.

National Association of School Nurses. (2006a). Position statement: Caseload assignments. Retrieved: February 5, 2008, from http://www.nasn.org/Default.aspx?tabid=209.

National Association of School Nurses. (2006b). Position statement: Emergency care plans for students with special health care needs. Retrieved: February 5, 2008, from http://www.nasn.org/Default.aspx?tabid=220.

National Association of School Nurses. (2004). Position statement: Medical services vs. health services in the school setting. Retrieved: February 5, 2008, from http://www.nasn.org/Default.aspx?tabid=229.

Neugut, A.I., Ghatak, A.T., & Miller, R.L. (2001). Anaphylaxis in the United States: An investigation into its epidemiology. *Archives of Internal Medicine*, 161(1), 15-21.

Nowak-Wegrzyn, A. (2003). Future approaches to food allergy. *Pediatrics*, 111(6), 1672-1680.

Olympia, R., Wan, E., & Avner, J.R. (2005). The preparedness of schools to respond to emergencies in children: A national survey of school nurses. *Pediatrics*, 116(6), e738-e745.

Primeau, M.N., Kagan, R., Joseph, L., Lim, H., Dufresne, C., Duffy, C., Prhcal, D., & Clarke, A. (2000). The psychological burden of peanut allergy as perceived by adults with peanut allergy and the parents of peanut-allergic children. *Clinical and Environmental Allergy*, 30, 1135-1143.

Salend, S. J. (2008). *Creating inclusive classrooms: Effective and reflective practices.* (6th Ed.). Upper Saddle, NJ: Merrill.

Sampson, H.A. (2004). Update on food allergy *Journal of Allergy and Clinical Immunology*, 113(5), 805-819.

Sampson, H.A., Mendelson, L., & Rosen, J.P. (1992). Fatal and near-fatal anaphylactic reactions to food in children and adolescents. *New England Journal of Medicine*, 327 (6), 380-384.

Sampson, M.A., Muñoz-Furlong, A., & Sicherer, S.H. (2006). Risk-taking and coping strategies of adolescents and young adults with food allergy. *Journal of Allergy and Clinical Immunology*, 117 (6). 1440-1445.

Sheetz, A.H., Goldman, P.G., Millett, K., Franks, J.C., McIntyre, C.L., Carroll, C.R., et al. (2004). Guidelines for managing life-threatening food allergies in Massachusetts schools. *Journal of School Health*. 74 (5):155-60.

Sicherer, S.H. (2002). Clinical update on peanut allergy. *Annals of Allergy, Asthma, and Immunology*. 88(4), 350-361.

Sicherer, S.H., & Malloy, T. (2005). The complete peanut allergy handbook: Everything you need to know to protect yourself and your child from the most deadly food allergy. New York: Berkley Books.

Sicherer, S.H., Furlong, T.J., DeSimone, J., & Sampson, H.A. (1999). Self-reported allergic reactions to peanut on commercial airliners. *Journal of Allergy and Clinical Immunology*, 104(1), 186-189.

Sicherer, S.H., Furlong, T.J., DeSimone, J., & Sampson, H.A. (2001). The US peanut and tree nut allergy registry: Characteristics of reactions in school and day care. *Journal of Pediatrics*, 138(4), 560-565.

Sicherer, S.H., Muñoz-Furlong, A., & Sampson, H.A. (2003). Prevalence of peanut and tree nut allergy in the United States determined by means of a random digit dial telephone survey: A 5-year follow up study. *Journal of Allergy and Clinical Immunology*, 112(6), 1203-1207.

Simons, F.E.R. (2004). First-aid treatment of anaphylaxis to food: focus on epinephrine. *Journal of Allergy and Clinical Immunology*, 113(5), 837-844.

Simons, F.E.R. (2008). Anaphylaxis. *Journal of Allergy and Clinical Immunology*, 121(2), 402-407.

Thies, K.M. (1999). Identifying the educational implications of chronic illness in school children. *Journal of School Health*, 69(10), 392-397.

Valenta, R. & Kraft, D. (2002). From allergen structure to new forms of allergen specific immunotherapy. *Current opinion in immunology*, 14(6), 18-27.

Weiss, C., Muñoz-Furlong, A., Furlong, T.J., & Arbit, J. (2004). Impact of Food Allergies on School Nursing Practice. *The Journal of School Nursing*, 20(5), 20-31.

INDEX

B

C

D

E

F

G

H

I

J

K

L

M

N

O

P

Q

R

S

T

U

V

W

X

Y